Surveying Northern
British Columbia

Surveying Northern British Columbia

A Photojournal of Frank Swannell

Jay Sherwood

Caitlin Press Inc.
Prince George, BC
2004

Copyright © Jay Sherwood 2004

All rights reserved. No part of this publication may be reproduced, stored in a retrieval system or transmitted, in any form or by any means, without the prior permission of the publisher, or in the case of photocopying or reprographic copying, a licence from Access Copyright, the Canadian Copyright Licensing Agency, 1 Yonge Street, Suite 1900, Toronto, Ontario, M5E 1E5, , 1-800-893-5777, info@accesscopyright.

Published by
Caitlin Press Inc.
Box 2387
Prince George, BC V2N 2S6

Design by Warren Clark Graphic Design
Typeset by Warren Clark Graphic Design
Maps by Warren Clark
Index by Kathy Plett
Cover photo BC Archives
All photos in this book, with the exception of those in "90 Years Later" are from the British Columbia Archives. Those in "90 Years Later" are the author's own.
Printed in Canada

Caitlin Press acknowledges the financial support from the government of Canada through the Book Publishing Industry Development Program and the Council for the Arts, and from the Province of British Columbia through the British Columbia Arts Council and the Book Publisher's Tax Credit through the Ministry of Provincial Revenue. Caitlin Press also wants to acknowledge the generous financial assistance of the Corporation of Land Surveyors of the Province of British Columbia.

Library and Archives Canada Cataloging in Publication

Sherwood, Jay, 1947-
Surveying Northern British Columbia : Photojournal of Frank Swannell, 1908-1914 /Jay Sherwood.

Includes bibliographical references and index.
ISBN 1-894759-05-2

1. British Columbia, Northern—Pictorial works.
2. British Columbia, Northern—History—20 th century—Sources.
3. Surveyors—British Columbia, Northern—Diaries.
4. Swannell, F.C. (Frank Cyril) II. Title.

FC3845.N67S53 2004 971.1'803 C2004-905138-5

Dedication

This book is dedicated with love to my family, and in honour of my parents.

This book is also dedicated to Gerry Smedley Andrews. Gerry articled for his surveying under Frank Swannell, and was Surveyor-General from 1951 to 1968. He provided invaluable information, and his support and enthusiasm for the book is greatly appreciated. I was privileged to get to know Gerry and spend many wonderful hours in his office at his home.

In addition, this book is dedicated to Mrs. Lil McIntosh who has been involved with the history of the Nechako Valley for many years. Through Lil I became a member of the Nechako Valley Historical Society where I first learned about Frank Swannell. Lil's knowledge of the history of the area and her moral support have been important for this book.

Preface

Land Surveying is a profession that uniquely combines academic knowledge of classic subjects such as astronomy, geometry, trigonometry and law, with practical skills such as line-cutting, measuring and note keeping.

This book is a tribute to a gifted land surveyor, Frank Swannell, who possessed this combination of knowledge and skills, and was in the right place at the right time to make a significant contribution to a young British Columbia. He also had the foresight to record his exploits in journals and photographs. These photographs were not just snapshots, but were masterfully composed and technically excellent. They are all the more remarkable given the extreme conditions under which they were taken.

Jay Sherwood has assembled a wonderful collection of these photographs and has skillfully woven them together with words to bring to life a remarkable man who lived in a romantic era of packhorses and unmapped wilderness.

The work of Frank Swannell and other early land surveyors was essential for the orderly settlement and development of the province of British Columbia. These men not only explored the wild spaces and laid out boundaries for pioneer homesteaders; they also assessed and reported on timber and agricultural potential, coal and mineral occurrences, transportation routes and a myriad of other details. All of this information was meticulously recorded in field books and later plotted in the form of beautifully crafted plans that are still referred to today when modern land surveyors retrace the boundaries of old District Lots and Sections.

The Corporation of Land Surveyors of the Province of British Columbia is proud to provide support for this book, and to recognize the accomplishments of one of our most respected members, Frank Cyril Swannell, PLS No. 75. The timing is especially fitting as we prepare to celebrate 100 years as a self-governing professional body in January, 2005.

David C. Bazett CLS, BCLS

President,
Corporation of Land Surveyors of the Province of British Columbia

Acknowledgments

There are many people who helped in the development of this book. Thank you to my wife for reading each chapter and providing feedback, and for your encouragement and patience. Thank you also to my two sons for their support.

The financial assistance of the British Columbia Land Surveyors Corporation is gratefully acknowledged. Bill McIntosh provided the first contact with the BCLS. Dave Bazett, President of the BCLS, and Rob Tupper, of the centennial committee, were instrumental in obtaining the Corporation's assistance. Robert Allen, of the historical committee, read the sections pertaining to surveying and provided his comments. Janice Henshaw efficiently and cheerfully handled the correspondence for the organization.

I would like to express my sincere thanks to Cynthia Wilson and Caitlin Press for publishing this book. I hope that this book supports your goal of making people more aware of the history of the BC Interior, and justifies your faith in this project.

Thank you to the staff of the British Columbia Archives for their assistance, particularly the people who cheerfully brought out the same boxes of material every time I was there. Janet Mason of the Base Mapping & Geomatic Services Branch spent a morning explaining the work of her department and allowed me to browse through the geographical record cards. Jeff Beddoes arranged for me to spend time at the Surveyor-General's office. Browsing through old field books, journals and photos took me back to a time almost a century ago when surveyors were busy making surveys throughout the province.

A special thanks to Gerry Smedley Andrews and Lil McIntosh who provided information and support over many years. Frank Swannell's son, Art, gave me an interview in 1986, and I am thankful for the tape that I have of our conversation. Finally, thank you to my friends in Vanderhoof, and to the people on my trip this summer, for your enthusiasm. It was wonderful to work on this book and be supported by so many people.

Jay Sherwood
August 2004

Table of Contents

Dedication *v*

Preface *vii*

Acknowledgments *viii*

Introduction *1*

Early Surveying in British Columbia *5*

1908 *11*

1909 *25*

1910 *43*

1911 *65*

1912 *85*

1913 *103*

1914 *127*

90 Years Later *153*

Appendix 1 – The people who surveyed with Frank Swannell *155*
Appendix 2 – List of British Columbia Archival photographs *157*

Sources Consulted *160*

Index *161*

Introduction

Frank Swannell was one of the most famous surveyors in British Columbia. The Provincial Archives of British Columbia (now the British Columbia Archives) noted in a brochure, "Swannell's particular distinction as a surveyor lay in his ability to survey large areas quickly and accurately while working under difficult conditions." Don Thompson, in *Men and Meridians*, a three-volume history of surveying in Canada, observed that Swannell's work provided "the first mapping and surveys of many areas of British

Running transit at Lac La Hache. In the spring of 1908 Swannell did some surveying around Lac La Hache while en route to the Nechako Valley.

Columbia which showed clearly and with accuracy the valleys, mountains, and other configurations of land."

Even more important for the history of British Columbia is Swannell's photographic collection. His journal entry for April 4, 1901 noted: "bo't camera." Soon afterwards the first pictures began to appear in his surveying journals. From 1901 onwards Swannell took a camera on virtually all of his surveying trips. His collection of over 5,000 photographs is housed in the B.C. Archives and provides an invaluable record of the history of the province. While Swannell was not a trained photographer, he had the photographer's eye. The composition of his photographs, his attention to detail, and his ability to have people act naturally enabled Swannell to take pictures that have been used in countless books about British Columbia.

Frank Swannell was born in Hamilton, Ontario on May 16, 1880. When he was a young boy his family moved to Toronto where he graduated from high school. In the summer after his graduation Swannell went west for the first time, working in the wheat fields of Manitoba. From 1897 to 1899 he took a two-year program in mining engineering at the University of Toronto and obtained a certificate in assaying. In the summer of 1898 Swannell traveled to British Columbia where he worked for the surveying firm of Drewry and Twigg in New Denver, a mining town in the West Kootenays. Drewry and Twigg specialized in mining surveys. After graduating in 1899 Swannell intended to go to the Yukon to work in mining in the Klondike gold fields, but by the time he reached Victoria Swannell realized that he did not have enough money to get to the Yukon. He obtained a surveying job with Gore, Burnet and Co., which soon became Gore and McGregor. Swannell articled to T.S. Gore and in 1903 received his Provincial Land Surveyor's licence (PLS #75). The following year he received his DLS (Dominion Land Surveyor's licence). Swannell continued to work with Gore and McGregor until 1908 when he decided to begin surveying on his own.

This photojournal celebrates the life of Frank Swannell, a pioneer surveyor, and covers the years from 1908 to 1914, seven seasons that Swannell surveyed in northern BC, making the first surveys of a large portion of this region. Leaving Victoria in the spring, Swannell and his survey crew would spend at least six months in the field. For most of the time they lived in tents, away from the comforts of civilization. They worked six days a week, and ten to twelve hours each day. Mosquitoes and black flies made summers uncomfortable. Spring often meant cool and rainy days, while in the late fall the men contended with cold weather and snow. In the remote areas of northern BC there were rivers to cross and mountains to climb, and Swannell did not know what conditions he would encounter. The men had to adapt to the situations they faced, and be determined to overcome any difficulties that they confronted. Yet by all accounts from people who knew him, it was a life that Swannell loved. He liked working outdoors and having the opportunity to travel to parts of the province that few people visited. Swannell enjoyed meeting the Native and non-Native people of these remote regions, and the camaraderie of the men on his surveying crews. Many of these people became life-long friends. Swannell found that he particularly enjoyed the exploration surveys that he did for the government, and he had a special talent for this work. These surveys, undertaken in remote, rugged conditions, established Swannell's reputation as one of the premier surveyors in the province and brought him national recognition. The maps produced from his surveys were so accurate and detailed that they were not supplanted until the development of better surveying equipment and more advanced technology like aerial photography.

This photojournal recognizes the significance of the pictures that Swannell took during these years for the history of British Columbia. This was a time of great change in the northern half of the province, and with his observant photographic eye he recorded the life and times of the region. Swannell's surveys took him to distant locations where he saw First Nations people who still followed their seasonal, self-sustaining way of life. He witnessed the decline of the Hudson's Bay Company's influence in the region. Swannell traveled the Yukon Telegraph Trail and many of the other famous trails in the region. He saw the end of the stagecoach era, viewed the heyday of the sternwheelers on the Fraser and Skeena waterways, and observed the coming of automobiles and the Grand Trunk Railroad. Swannell met many of the early settlers, traveled rivers that have now been altered by hydroelectric dams, and surveyed valleys that even today remain remote.

For his pictures Swannell usually tried to have people be as natural as possible. He was successful in this respect considering that most of these people had probably never or seldom been photographed. In many of the pictures, particularly of the Natives, the people are sitting on the ground, and Swannell often took his picture at the same level. Occasionally Swannell took a formal portrait or a staged photograph. George Copley, Swannell's close friend who worked on his survey crews from 1909 to 1914, sometimes liked to strike a dramatic pose for a picture. Swannell took all of the photographs except for the ones in which he appeared. These were taken by various members of his survey crew. Despite the conditions in which Swannell worked and traveled, he was able to take pictures that were consistently sharp and clear, and his photographs draw the viewer into the scene. Swannell developed his own photographs, and some journal references mention developing pictures while he was out on his field surveys.

Swannell lived a life of rugged, romantic adventure, but the surveys that he made were part of the changes that were irrevocably altering the life of the region. The field journals that Swannell kept, which are housed in the British Columbia Archives, and the photographs that he took show us what it was like to be a surveyor almost one hundred years ago and provide a window into British Columbia's history. This book is a celebration of Swannell's life and recognition of the importance of his photographs to the history of this province.

A note on terminology -

Swannell uses the word Indian to describe the First Nations people that he met. This was the name that was commonly used at the time. He also uses Siwash, another word that was used for the First Nations people. Siwash was sometimes used as a racial, derogatory term depending on the person who used it and the tone in which it was used. Throughout his journals Swannell writes respectfully of the First Nations people he met, and by all accounts he had a good relationship with them, so the term Siwash was used simply as a variation for Indian.

In the early 20th century surveyors used feet and miles for their measurements. A mile is 1.6 km. A foot is 30 cm.

Early Surveying in British Columbia

Frank Swannell at Takla Lake
Swannell is surveying at a triangulation station on Takla Lake. His survey helper and cook, Nep Yuen, is using a plumb bob to help Swannell locate the position where the transit should be set. Swannell had set this triangulation station in 1912 but found it underwater when he returned in 1913.

Surveyors played an important role in North America, particularly in the 19th and early 20th century settlement of the western part of the continent. As the First Nations people in the United States were displaced and put onto reservations, settlers moved into the American Midwest and Great Plains and began farming the land. Surveyors divided the land into an orderly grid known as the township and range system, consisting of township lines that ran east and west at six-mile intervals, and range lines which ran north and south every six miles. This produced a system of six-mile squares. Each six-mile square was divided into 36 one-mile squares called sections. The section, which totalled 640 acres, was divided into quarters of 160 acres. Since the north-south range lines converged towards the North Pole, adjustments (called correction lines) were made at intervals along the township lines. This became particularly important in the northern latitudes of Canada where the north south lines converged more noticeably. The township and range grid, with its unit of one-mile squares, produced the famous checkerboard pattern of land settlement which became the hallmark of western North America.

The surveyor's equipment was simple, yet it enabled him to divide the land with a high degree of accuracy. The transit, also known as theodolite, was used to measure angles. A 66-foot chain, divided into 100 links, measured distance, with 80 chains equalling one mile. Wooden, and later, iron posts were set into the ground every half mile along the survey lines. A survey crew usually consisted of four people: two to measure the distances, one to use the transit, and the surveyor who recorded all the distances and angles, and ensured that the work was being performed accurately. When the land being surveyed was covered with trees, more men were hired to cut the survey line with axes.

In addition to surveying the land, the surveyor made a record of the country through which he traveled. He noted the topography, soil, water drainages, and potential use of the land, thus providing invaluable information to both government and to the settlers.

The 1864 Homestead Act in the United States gave settlers 160 acres of land, provided they fulfill certain conditions for farming it. This Act was a major impetus for settlement of the United States west of the Mississippi River. From the end of the Civil War in 1865 to 1900, surveyors were busy surveying land across the American west to the Pacific Ocean. The township and range system of dividing and marking the land proved so successful in the United States that it was even used in the Rocky Mountains.

Surveying in western Canada followed patterns established in the United States. After Canada became a country in 1867, the federal government took over ownership of the western lands from the Hudson's Bay Company in 1869. The federal government assumed responsibility for the Northwest Territories, the vast area of land which later included the provinces of Manitoba, Saskatchewan, Alberta, and the northern territories. With a few minor modifications, the federal government adopted the township and range system used in the United States. Free land was also offered to settlers to encourage them to settle on the Canadian Prairies. Throughout the last three decades of the 19th century, settlers came to the Prairies in increasing numbers, particularly after the completion of the Canadian Pacific Railroad in 1885, and as good farmland in the United States became scarce. The federal government employed many surveyors to divide the Prairies into townships, ranges, and sections, producing a radical change in the shape, use, and population of the land. By the end of the 19th century over 81 million acres of land had been surveyed between the Manitoba border and the Rocky Mountains.

However, in Canada's westernmost province surveying and land use followed a different direction, due to terrain, funds and politics. British Columbia became a province of Canada in 1871 with the promise of a transcontinental railroad to link it with the rest of the country. The Railway Belt, a forty mile strip of land on both sides of the Canadian Pacific Railroad's route through British Columbia, and the Peace River Block, an area of land in the northeastern part of the province, were given to the federal government as part of the deal for construction of the railroad. The British Columbia provincial government had control over the administration of the rest of its land, but it was also responsible for the surveying and administration expenses.

Surveying in British Columbia occurred in response to the topography of the province and the demand for land by people and companies. In most areas of the province agriculture was not the main use of the land, and the mountainous terrain did not lend itself to the township and range system. Settlement was scattered, not in the orderly patterns established on the Prairies or in the American West. The British Columbia government had limited funds to spend on surveying land, and it could not predict where and how settlement would evolve, so it did not develop an orderly surveying system. The province was divided into several land districts. Within a district each parcel of land surveyed received a

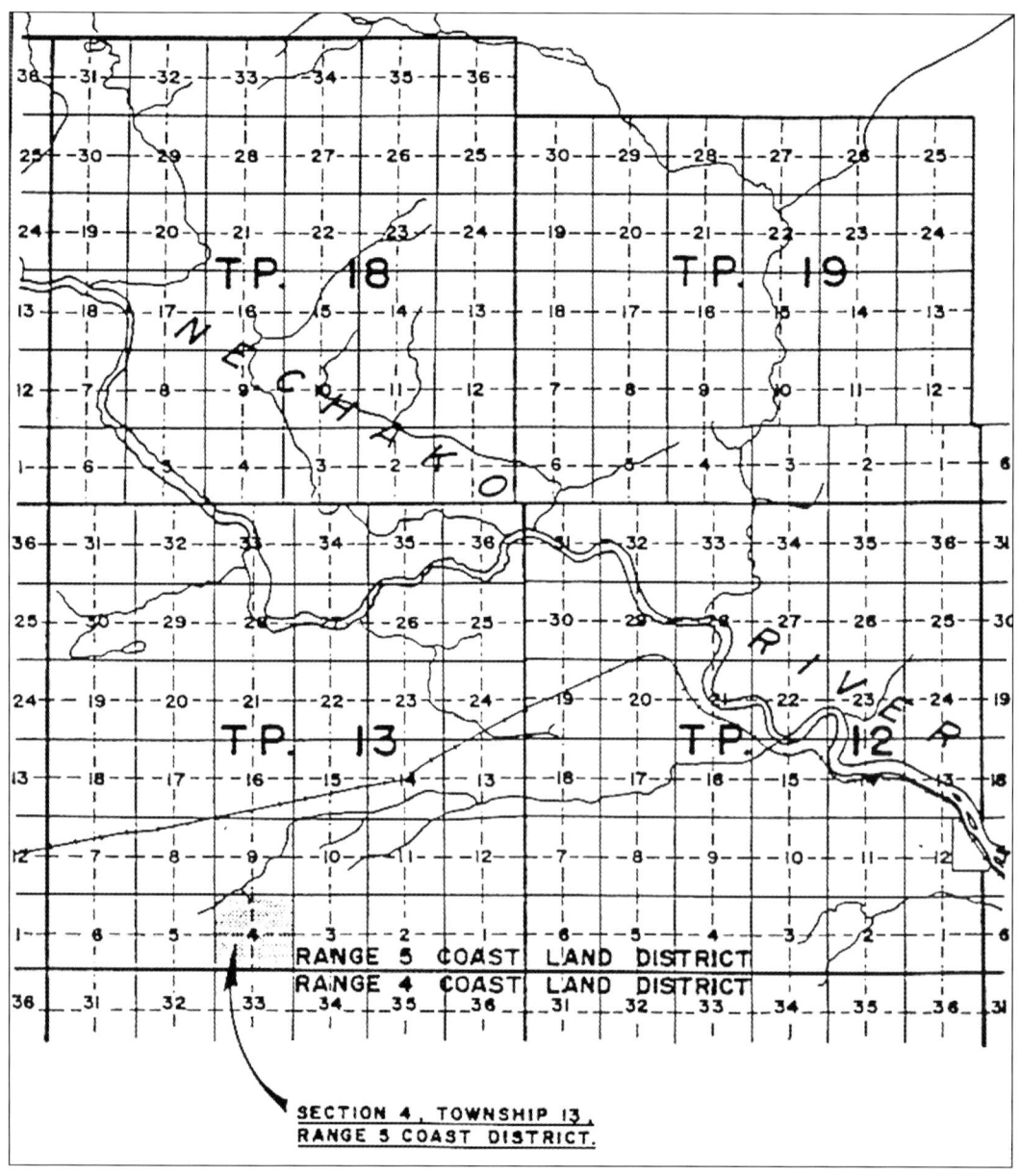

■ An example of a township and range system in British Columbia. Township 13 was one of the townships in the Nechako Valley that Frank Swannell surveyed in 1908. Source: Taylor, W.A. Crown Lands: a History Of Survey Systems.

lot number. Lots were numbered consecutively within the district regardless of where they were located. The size and shape of lots varied, depending on the use of the land and the local terrain. A patchwork lot system, often unconnected with other lots, developed. The provincial government surveyed only a minimal part of the province, while many areas, particularly in central and northern British Columbia, remained largely unsurveyed and unexplored by the beginning of the twentieth century.

Tom Kains, the Surveyor-General for the British Columbia government from 1891 to 1898, recognized that there was a need for some type of accurate surveying system in the province that would link together the lots which had already been surveyed, and which could be used to join with future surveys in British Columbia. Kains realized that the township and range system was not practical for most of the province. Instead he looked to the surveying systems being used in New Zealand, which had terrain similar to British Columbia, and to Europe.

Kains proposed a triangulation network with control points. Triangulation is a type of surveying where control points (survey monuments) are set, usually on a high point of land, or in an open location in a valley which has a good view of the surrounding area. A short baseline (also known as a subtense line) is established and the distance is measured along this line. From the two ends of the line angles are read to the control points that can be seen at each location. If two angles and a distance are known it is possible to calculate the remaining angle and distances through trigonometry. New control points could then be set further away and measured from the control points that had already been established. To provide more distances new short baselines could be set up and tied into the established control points.

This triangulation system would enable the government to set up a survey network that accurately covered a large area in a relatively short time at a reasonable cost. The lots surveyed in an area could be tied into this triangulation system. By measuring the vertical angle between control points, triangulation could also be used to develop a topographic map of the region being surveyed. Kains supported a partial use of the township and range system for locations where there was a large area of land that would be used mainly for agriculture. These areas would still be tied in to the triangulation network that he proposed. While Kains was Surveyor-General, he

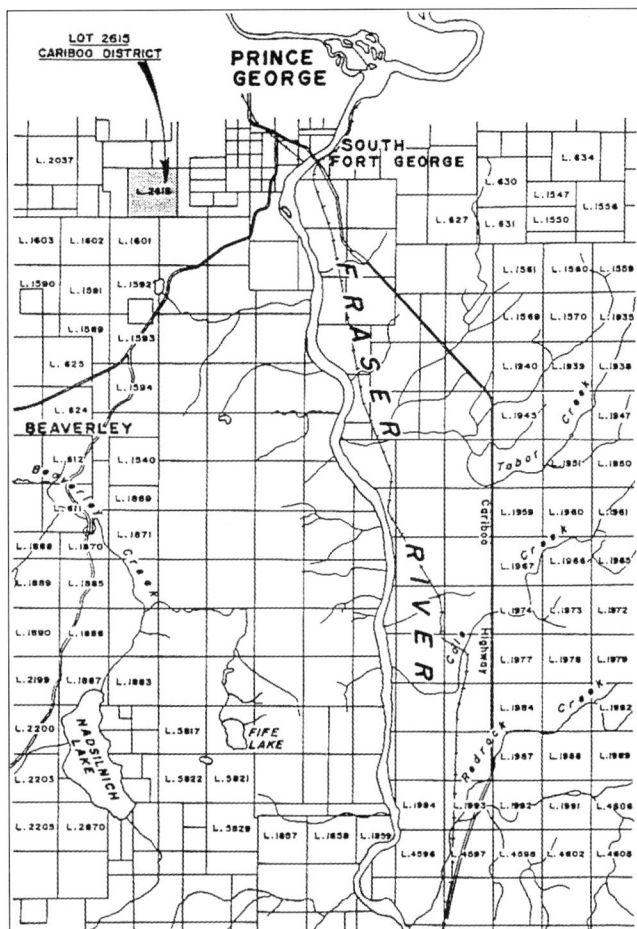

■ This is a lot system in the Cariboo District near Prince George. Source: Taylor: Crown Lands

began implementation of his proposal through the establishment of a series of Mineral Monuments in the Kootenays. Kains instructed the surveyors in the region to tie in their lot surveys to these Mineral Monuments whenever possible. One of the surveyors involved with these Mineral Monuments was W.S. Drewry, for whom Frank Swannell worked in the summer of 1898. After Kains left as Surveyor-General in 1898 his triangulation network was not continued. However, Kain's progressive ideas would later be implemented.

In his government report for 1912, G.H. Dawson, the Surveyor-General, bemoaned the problems the provincial government encountered in the development of surveying in British Columbia.

Comparison is frequently made between the surveys of this Province and those of the other Western Provinces, and while the superiority of the latter is abundantly evident, the reason for the inferiority of the former is often lost sight of. The conditions in this Province differed so materially from those of other Provinces that a different method of meeting them had to be adopted…It is not suggested that this Province attempt to follow in the wake of the other Western Provinces and adopt the highly scientific system of the latter, nor is it suggested that radical changes be made in the spirit and wording of the Statues. The Province has reached its present state of development without such a system, and it is too late to change. As, however, the lack of available funds which has up to recent years crippled this Branch is now at an end, it is possible to improve the field-work to such an extent as to ensure results in the future which will, considering the limitations imposed by local conditions, be comparable with those achieved in other Provinces. The difference between existing conditions as regards land surveys in this province and the other Provinces is due to the mountainous character of the former, the variety of its resources, and the fact that this Province alone surveyed and administered its Crown lands.

The Surveyor-General further lamented that, "The Dominion government, with its relatively unlimited resources, undertook the survey of the vast prairie area, which presented no obstacle to the adoption of any system, and through the greater part of which transportation by wagon was easy. The country was blocked out by surveyors at day rates…the whole being completed long in advance of settlement." By contrast, Dawson wrote, "the Crown Colonies of Vancouver Island and British Columbia were confronted, with an empty Treasury, with the problem of surveying a territory almost as large, with physical features which made a systematic survey of the whole practically impossible, and which was, for the most part, covered with a growth of timber which rendered the country absolutely inaccessible, even for packhorses, and necessitated supplies for survey parties being carried on men's backs. Dawson also noted that, "Surveys were carried out spasmodically by the

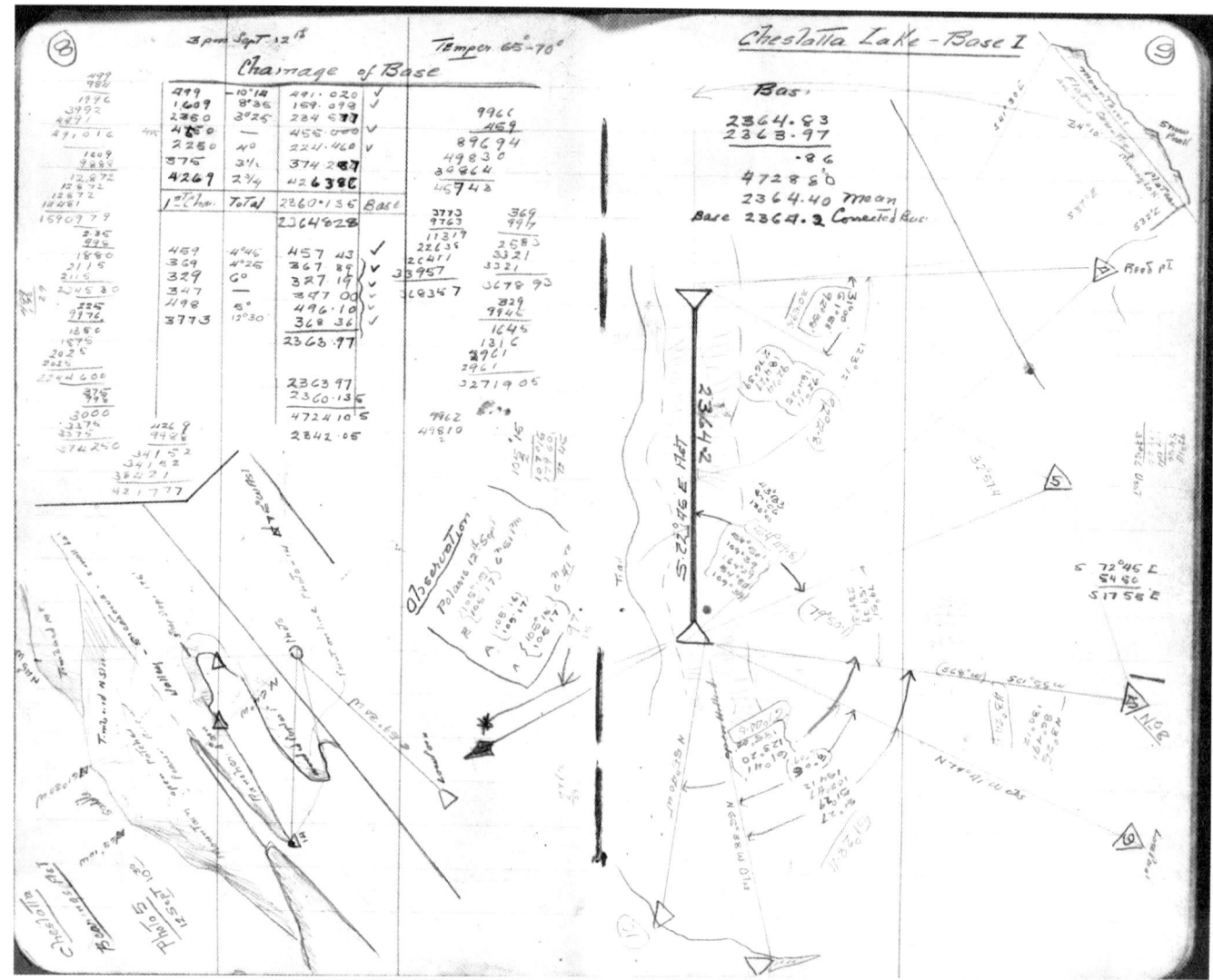

■ From Swannell's triangulation of Cheslatta Lake in 1910, is typical of his field notes. The distances for a baseline are recorded, along with some of the angles measured by the transit, and some of his calculations.

[Provincial] Government, and that no systematic effort was made to survey the Province as a whole is proved by the expenditure on surveys each year."

Surveying in British Columbia changed dramatically early in the 20th century. In 1903 Richard "Glad-Hand Dick" McBride was elected Premier of the province. At the time of his election, the public debt stood at over $12,000,000 and the credit of the government was exhausted. The Minister of Finance, R.G. Tatlow, brought order into government finances, and by 1905 had produced a slight surplus, only the second real one in British Columbia since Confederation. This was the beginning of several years of prosperity in British Columbia. Timber resources in the United States were diminishing and people turned to the seemingly boundless timber in British Columbia's forests. The fishing industry was booming. The mining industry in the Kootenays was continuing to flourish. The orchard industry in the Okanagan was expanding. The second transcontinental railroad, the Grand Trunk Pacific, was being constructed. People believed that this railroad would open the northern part of the province to development, particularly agriculture, and there was a land boom in the region. The Canadian National Railroad was also being built, and there was speculation that other regional railroads would be developed. This economic development spurred the need for land to be surveyed and for the provincial government to spend more money on surveying.

In 1905-6 the provincial government spent less than $7,000 on surveying, but by 1910-11 the amount was $448,885. From 1909 to 1913 an average of over 4% of the provincial budget was spent on surveying, reaching a peak of almost 6% in 1911. In the years since then the provincial average has been less than 1%. The number of lots surveyed jumped from 1,830 in 1907 to 7,312 in 1911. The number of acres surveyed increased from 1,189,428 in 1909 to 3,226,610 in 1911. Of this land surveyed 1,352,809 acres were purchased, 686,909 went to timber limits; 120,938 were used for coal licences, and 948,644 acres were surveyed by the government to be ready for future sales. The Surveyor-General's department expanded from 15 to 41 people in 1911 to keep up with the number of land surveys being processed. A sub-branch was established to produce maps. Two inspectors were appointed to check on the surveyors' work in the field.

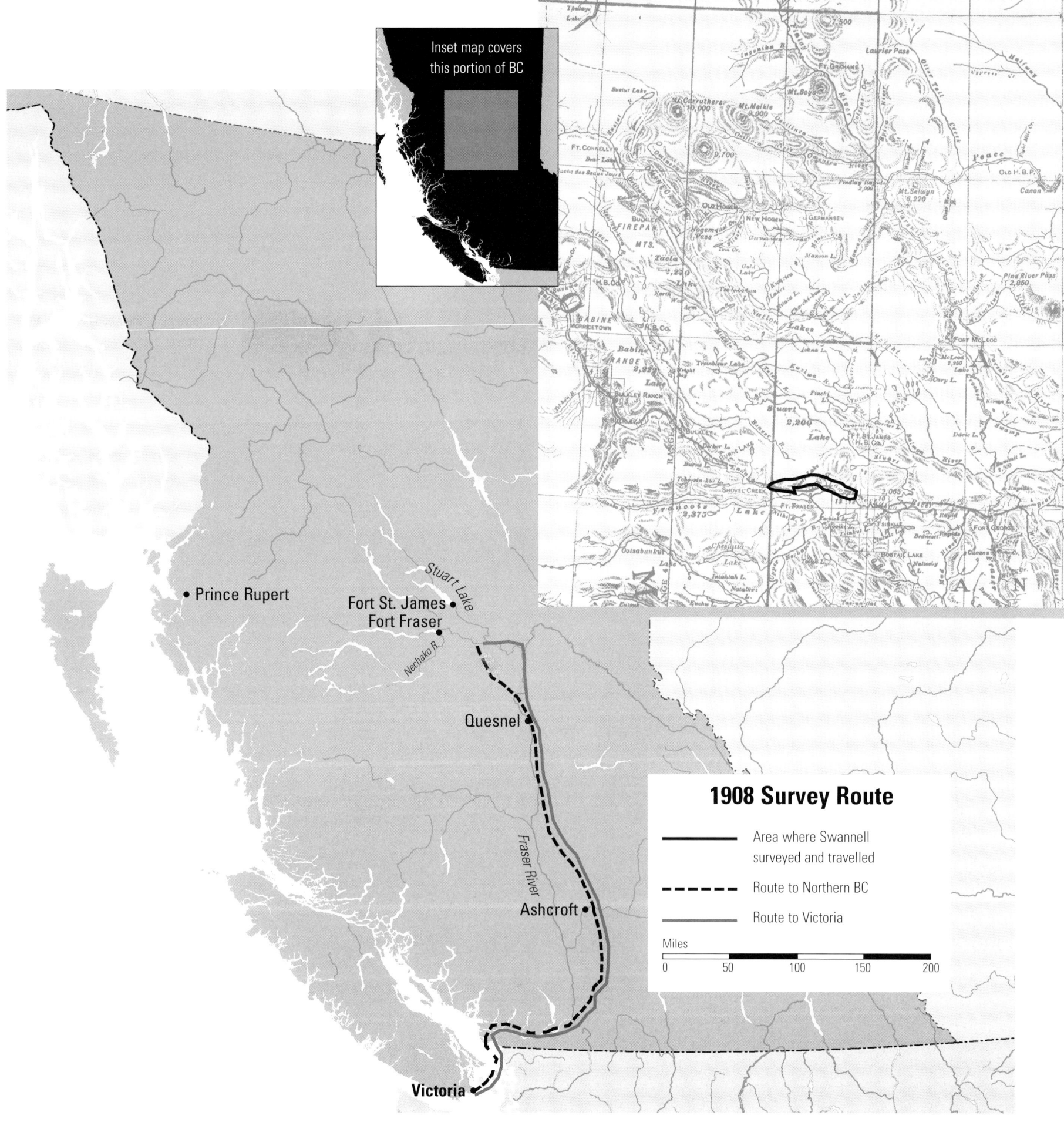

1908

In 1908 Swannell decided to leave Gore (LS) and McGregor (PLS #1), the surveying company for which he had worked since arriving in British Columbia in 1899. He had been a registered surveyor since 1903, and the economic growth in the province was providing more work for surveyors. Swannell went into partnership with another surveyor, A.I. Robertson (BCLS #24), and they received a contract from the provincial government for surveys in the Nechako Valley. This was the beginning of seven seasons of surveying in northern British Columbia for Swannell.

In 1908 the provincial government's primary interest in the northern part of the province focused on the

■ Swannell survey notes: "Muskeg, no bottom with 9', picket." The men have all of their body covered to protect themselves from mosquitoes and are working in waist-deep water.

■ Survey camp opposite Quesnel. Swannell's survey crew is on the west bank of the Fraser River. They're packing up their gear and getting ready to start up the Yukon Telegraph Trail to the Nechako Valley. The ferry crossing was at the north end of Quesnel, so the main part of the town would be to the right, out of the picture.

Grand Trunk Pacific, which was constructing the second transcontinental railway in Canada. The railroad entered British Columbia through Yellowhead Pass in the Rocky Mountains west of Jasper. Its route followed the upper part of the Fraser River until its junction with the Nechako River at Fort George. Then the railroad traveled west up the Nechako and Endako Rivers, over a low divide, and down the Bulkley and Skeena Rivers to the Pacific Ocean at Prince Rupert.

The British Columbia government believed that the railroad would bring settlers to this part of the province and make the natural resources of the area more accessible. There seemed to be extensive timber and considerable mineral potential, while some valleys appeared to be suitable for agriculture. The provincial government hoped that the railway would spur the development of

■ Indian burial site at Graveyard Lake. These graves are from a smallpox epidemic, either in 1862 from the Cariboo gold miners, or from the Collins Overland Telegraph men in 1865-66.

■ Making flapjacks for the survey crew. This was a Sunday meal when there was enough time to cook pancakes for all the men. During the other six days of the week the men had a quick breakfast and got to work as soon as possible.

agriculture, which had been of only minor importance in British Columbia. Almost all the arable land on the Prairies had been homesteaded, so the government thought that land-hungry settlers would be attracted to what was being billed as the last agricultural frontier in Canada. Conditions would be harsher than on the Prairies and work would be needed to clear the land of brush, but there was optimism that the land had great potential. The British Columbia government anticipated that the primary area for agriculture would be located in the Nechako Valley.

By 1908 the Grand Trunk Pacific had surveyors locating the railway line in the province, and construction work had started. Settlers were arriving in the Nechako Valley, looking for agricultural land. The provincial government wanted to ensure that there was orderly development and that they received their land sales revenue. Although the traditional method of surveying an area in British Columbia was to divide the land into lots within a district, the Nechako Valley was large and flat enough to be surveyed using the township and range system. The British Columbia government decided that surveying by this method would facilitate the development of agriculture since farmers were used to this land pattern. In the 1890s the provincial government had surveyed a couple of townships in the Nechako Valley, and in 1906 and 1907 J.H. Gray (LS) had surveyed a few more. However, the large influx of settlers who wanted to take up agricultural land in the Nechako Valley pressured the provincial government to finish surveying the area as quickly as possible. Swannell and Robertson received one of the government contracts

■ Tsinkut Falls. This waterfall is located on Tsinkut Creek shortly before it empties into the Nechako River. Swannell had a camp along this creek for part of the summer.

Indian axemen. This is a staged photograph. The Natives working on the surveying crews are trying to look as fierce as possible with their weapons.

to survey the Nechako Valley, dividing it into one-mile square sections using the township and range system.

Swannell, with one surveying crew, left Victoria by boat for Vancouver on April 28. The next day they departed Vancouver by Canadian Pacific train for Ashcroft where they switched to horse transportation. During the next three weeks Swannell's crew surveyed some lots in the Cariboo. Robertson left Victoria on May 19 with two surveying crews and joined Swannell in the Cariboo. The surveyors traveled to Soda Creek where, on May 24, they boarded the sternwheeler *Charlotte* for an eleven-hour trip up the Fraser River to Quesnel. The *Charlotte*, which was built in 1896, operated between Soda Creek and Quesnel. In 1908 it was the only sternwheeler on the upper Fraser River.

Quesnel was the last sizeable town in the region, so Swannell and Robertson remained there a few days outfitting the survey crews, shoeing horses, hiring packers, purchasing supplies, and taking care of finances. From 1908 to 1913 Quesnel was Swannell's supply and financial cen-

ter. Among the businesses there was a Hudson's Bay store, a large grocery store operated by John Fraser, the MLA for the region, two banks, two hotels, a telegraph office, and an office for the sternwheelers. The *Cariboo Observer*, Quesnel's weekly newspaper that began in late 1908, usually noted the arrival of Swannell and his survey crew in town during the spring and late fall.

When these preparations were completed, Swannell arranged for the crews and equipment to be ferried across the Fraser River, and the ten-day journey to the Nechako Valley began. From Quesnel the route to the Nechako Valley proceeded northwest along the Yukon Telegraph Trail. This route had been constructed when the federal government built a telegraph line to provide communication to the gold mining communities of the Klondike. The trail had been built for maintenance of the telegraph line so it was rough and primarily suited for horse travel. Wagons were able to follow the route, although it was a slow, difficult journey.

The trip was uneventful except that "one bald-face

■ *SS Charlotte*. The *Charlotte* was the only sternwheeler operating on the upper Fraser River in 1908. It traveled between Soda Creek and Quesnel. Some of the cordwood for the boiler is stacked in the front. The gang plank from the sternwheeler to the shore is visible.

cayuse plays out, and later died." Along the route the surveyors camped one night beside the Yukon Telegraph Cabin at Bobtail Lake, and they observed the Native graves at Graveyard Lake. On June 6 they arrived in the Nechako Valley and established their first camp.

As head of the surveying outfit, Swannell had several responsibilities. He and Robertson shared the supervision of the surveying. Swannell had to hire the men that he needed for the survey crews. During their surveying in the Nechako Valley Swannell employed several Natives and settlers temporarily for a variety of tasks, and some of them worked for him again in the following years. Working on the survey crew provided a source of cash for some of the settlers while they were getting their farm established. Swannell also handled the finances. Most of his supplies were ordered on credit through the Hudson's Bay Company posts in Quesnel and at Fort Fraser. The Surveyor-General's office would either pay an advance to the Hudson's Bay Company or issue a letter guaranteeing payment to a designated amount of money. Although Swannell was in a remote area, he was close to the Yukon Telegraph Trail. Telegraphs were the high-speed communication of the early 20th century. Swannell often went to the Stoney Creek Telegraph cabin to pick up and send messages. On July 13 Swannell wrote, "Wire Lands Dept. about extending surveys until September." "Robertson & self at Telegraph Office all day," Swannell recorded on July 19. The journal entry for August 16 mentioned, "Receive telegram from Edgar Abbott – is bringing by canoe from Quesnel 800 lbs. flour, 3 cases milk." In September the Surveyor-General sent a telegram asking him to survey part of

Indian girls at Nulki Lake The girls are picking berries, probably saskatoons.

another township in the Nechako Valley. There was also mail service. Swannell wrote on June 6 that he met Karl Gauss with two packhorses, "His Majesties Mail."

Besides the surveying, Swannell's major responsibility was the logistics of feeding about twenty people. Since there were only a few settlers in the Nechako Valley, there was very little food that could be purchased locally, and the surveyors had little success in hunting. Before leaving Quesnel Swannell had ordered a large supply of food, but it took over three weeks for the supplies to arrive by pack train, and Swannell had to send some men to the Hudson's Bay post at Fort Fraser and the Native reserve at Stoney Creek to buy any food available. On June 29 Swannell sent Henry Deschamps to Stoney Creek for more supplies, but he returned the following day with only some yeast cake. July 5 – "Henry

■ Jenny Blench and her father, Tom. Jenny Blench was the first non-Native girl to settle in the Nechako Valley. In 1909 her father, Tom, received the contract to transport mail between the Nechako Valley and Quesnel. Jenny usually accompanied him on these trips. The *Cariboo Observer*, in its May 29, 1909 edition, reported that she christened the new sternwheeler, *Nechacco* when she was in Quesnel. In the fall of 1909, when she went to school in Kamloops, the *Cariboo Observer* noted that she rode a horse to Ashcroft.

■ Stoney Creek Indian Church.

D. to Stoney Creek for *more* grub." One of the settlers, Jack Charleson, was able to supply beef, and Swannell made several purchases from him. On August 1 Swannell noted that he gave Collins, at the Hudson's Bay post in Quesnel, an order to pay Charleson $775 for the beef that he had supplied. Despite the large amount of supplies brought in from Quesnel, Swannell found that sufficient food for the surveyors was a continual worry, and he found it necessary to make more purchases locally as well as ordering more food from Quesnel. On September 1 Swannell "gave Indian Joe order for 3000 lbs by canoe from Quesnel." In the fall, when locally grown food became available, Swannell ordered from Charles Constantineau, a French-Canadian prospector and settler, 100 pounds of potatoes, 50 pounds of turnips, 50 pounds of beets, 15 pounds of carrots, and two geese.

The process of dividing the Nechako Valley into 640-acre sections was simple – in theory. The survey lines ran north-south and east-west. Every six miles an east-west township line was established and a north-south range line was surveyed. Each mile a surveying post was set so that the area consisted of numerous one-mile squares. A post was also set every half-mile so that

■ Swannell's surveying party at Fraser Lake. Swannell is in the front at the left by one of the transits.

each section could be divided into quarters (160 acres) in the future. Occasionally observations of Polaris (the North Star) were taken to check that the surveying lines followed the compass directions. Swannell and Robertson's three surveying crews worked together, operating from a common base camp. Swannell supervised one crew; Robertson a second, with R.P. Bishop (BCLS#73) as his assistant, while R.M. Benson, a surveyor they hired, was in charge of the third. Swannell's field notes recorded the daily progress of the three crews in the Nechako Valley surveys.

This is a page from one of the surveys that Swannell

■ Surveyors taking sights at Eyrie triangulation station. Swannell is operating the transit.

Vital's ferry, Cataline's train. This is the pack train of Cataline, the famous packer of northern BC, being ferried across the Nechako River. Vital's ferry was located on the Nechako River near Fort Fraser, close to the intersection of the Yukon Telegraph Trail and the First Nations trail that ran from Fort St. James to Cheslatta and down to the Grease Trail. Vital LaForce was a French-Canadian prospector who discovered gold in the Omineca at Vital Creek. He also found gold on the Finlay River at a place known as Vital's Bar. In 1906, during the latter years of his life, LaForce established this ferry on the Nechako River.

ran in the Nechako Valley. The survey is read from the bottom of the page to the top. Swannell has sketched a swamp that was near the line, and noted large poplar and spruce trees in the middle section.

Several factors prevented the surveying from going smoothly and kept life interesting for the men. Since the surveying lines went north-south and east-west, the surveyors had to follow the terrain, and the axemen had to chop out the trees and brush that were along the line. In most places the timber was not too thick in the Nechako Valley, but at times progress was slow because a lot of clearing was necessary. On June 14 Swannell noted: "heavy cutting on N. Bdy Sec. 13"; on July 7 "heavy chopping all day." Swannell observed "bad country N. Bdy 36 Tp. XIV" on July 11. During the summer there were several injuries to the men who cleared the line. "Driver cuts leg" Swannell wrote on June 17. The entry for July 20 recorded that "Gilbert Forbes chops himself and is bro't to camp by McDougall." "Wilkins cuts his ankle" Swannell noted on August 3. On September 14 Swannell wrote that "Jim cuts foot" and the following day he recorded that, "Archie cuts foot very badly. Put in three stitches babiche, drawn thru with pliers, on a darning needle." Babiche are strips of rawhide or sinew.

At times the surveying lines went through swamps and muskeg. On June 25 Swannell failed to connect one of his survey lines because he became mired in a swamp, and the same day he surveyed "E bdy Sect 19 in deep muskeg." The following day Robertson "ran into large swamp at 9:30 and got thru at 3." Bishop and Swannell ran Section 30 through a swamp.

Mistakes by the surveyors caused difficulties occasionally. On August 22 Swannell wrote that "Benson's lines haven't closed at all…E. Bdy 28 T. 1 post 1 chain set too far…Find Bishop broke chain in 2 places & mending left out a piece a few links long. Several lines have to be re-run entirely." The next day, while "making correction surveys Wilkins loses clamping screw of tran-

sit". Two days later Swannell found that the "wrong angle had been turned off baseline. Hell to pay generally." On September 4 "Bishop rechains [measures] E. Bdy 19 – finds and corrects 2.95 chainage error" [approximately 200 feet], while later in the month Swannell recorded that an error of "5 chains [approximately 330 feet] by Bishop causes running of new line." One day Swannell's crew had to quit early because they had lost the magnifying glass used to read the angles on the transit.

Throughout the summer Swannell and his men were often bothered by flies and mosquitoes-a common scourge of surveyors throughout northern Canada. June 14 – "mosquitoes bad"; July 2 – black flies very bad all day." There were similar entries for July 13, 14, and 15. On July 23 Swannell noted "mosquitoes bad in morning." Windy days provided a welcome relief from the insects. June 22 – "windy and cool, fairly free from flies"; July 9 – "high wind and *no flies* in the evening." Not until mid-August did the insects become less bothersome. Swannell's entry for the 15th noted with relief, "Hardly any mosquitoes. Smudge not needed at lunch for first time."

Swannell's field notes regarding the weather indicated a potential problem for agriculture in the Nechako Valley – the possibility of frost throughout the summer. June 23 – "frost – settlers potatoes frozen." Two days later there was another frost. On July 12 there was a heavy frost, and later in the same month there was a heavy frost that "put the final crimp into Dad Hill's potatoes. Weather exceptionally cold for July." By August 10 there was a "heavy frost, ice on water in bucket." On August 27 there was a heavy rain and a hailstorm. Swannell's daily record of the weather often noted rain while warm weather or "fine days" are seldom mentioned.

With three crews working, the surveyors completed two to three miles of line almost every day. In order to avoid long walks to their surveying, camp had to be moved frequently. A campsite had to be centrally located for the area they were surveying, provide adequate feed for the horses, and have a source of sufficient water. Then all of the equipment, supplies, food, and personal belongings had to be moved. Several times Swannell employed First Nations packers from the Stoney Creek Reserve to help move camp. Sometimes life in camp did not function smoothly. On June 17 Swannell noted that three horses had gone missing, and on September 12 he wrote that "Buckskin, Darkey & Bay Mare lost – bro't back by Indian David." Swannell recorded that he "got turned round going out" and Johnny got lost on July 4. When they were moving camp on August 10 the horses were late and didn't make it to the new site so they weren't able to pitch any tents that night. In his August 12 entry Swannell wrote, "Charlie [Hemeyer] getting in wood – horse wouldn't pull. Charlie sells horse to Jean Paul." On another day "Transit fell off horse en route – no apparent damage," Swannell noted.

Generally the men were healthy, with only an occasional note of a person in camp sick or with an injury. One serious illness occurred in late August when one of the men became ill with stomach trouble. The closest doctor was in Quesnel so "Macfie arranges with Indian Donald to take Henry by canoe to Quesnel. Gets canoe to Tsinkut mouth at midnight. Indian Donald to get $3.00 per day for self and canoe. Hill, Clarke and I pack Henry Driver down to Tsinkut mouth."

Since the surveyors were paid by the government based on the number of sections surveyed the men worked about twelve hours six days a week throughout the summer. There were several days where it rained sufficiently to keep the men in camp for all or part of the day, and on occasion everyone helped move camp. A heavy rainstorm during the morning of June 15 forced the men to quit work. Four days later the men quit at 2 p.m. in a downpour. Swannell's entry for July 8 noted, "Rain in morning, and late breakfast, but started out. Rain all day – all hands in early – soaked to the skin." Once in a while the men had a respite. July 17 – "Move to Nulki Lake near Charlesons cabin. Take photos at Stoney Creek village. Fine swim in Nulki Lake." Stoney Creek, about ten miles south of Vanderhoof, is the site of a Carrier First Nations reserve. The Carrier have several reserves in the Nechako and Stuart River watersheds.

Swannell observed the settlers who were beginning to arrive in the Nechako Valley. July 17 – "Three

Nechako land hunting outfits camped at Telegraph Creek." Two days later he met a "Swede and wife on trail with baby and small express wagon." One week later an "incoming settler with wagon wants land surveyed." Swannell tried to maintain good relations with the people of the area and when he had sufficient food he would invite some of them to have dinner with the surveyors.

In late July Robertson left for Quesnel taking with him sixteen field books of notes, the record of their Nechako Valley township surveys that had been completed. On the last day of July Swannell noted a record day with the three crews surveying a total of six miles. By early September Swannell's crews had completed the surveys they were supposed to make in the Nechako Valley during that summer. Some of the men were laid off and left, while Swannell took about ten people with him to Fraser Lake. Swannell and J.H. Gray (LS), a surveyor who also had a crew working on a government contract in the area, agreed to a division of the remaining work.

From early September to mid-October Swannell's crew conducted a triangulation survey of Fraser Lake and tied it into the Nechako survey. In the early 1890s, the 54º latitude line in the Nechako Valley had been established while the Native reserve at Stoney Creek was being surveyed. The townships surveyed in the Nechako Valley were tied to the 54º latitude line. In his journal entry for June 21 Swannell recorded that he had surveyed one mile west on the 54th parallel. The triangulation of Fraser Lake would enable the provincial government to know the location, size, and shape of the lake very accurately.

The surveys around Fraser Lake proceeded with few difficulties. From one of the settlers Swannell purchased a bateau, a canoe-shaped boat that could hold several people. This made it easier to conduct their survey around the lake except for days when high winds made it difficult to paddle. They were able to purchase some fresh vegetables from the settlers and fish from the Natives, so it was easier to keep the men fed. The only potential conflict occurred while Swannell's crew camped on the Stellako Reserve at the west end of Fraser Lake. "Siwash dog runs away with slab of bacon & is shot by Sharkey [the cook]. As this occurs on the Indian reserve and is strictly illegal, dog is cremated, leaving no

■ Travelling by scow from Fraser Lake to Soda Creek. The Fraser River is starting to freeze and is filled with slush ice.

evidence." There were a few level areas along the lake where settlers were establishing farms, and Swannell surveyed lots for them. These lots were surveyed in the traditional method, although they were connected to his survey of the lake. Swannell's crew also extended their triangulation survey to include some small lakes located south of Fraser Lake.

By October 19 Swannell had completed his surveys in the Fraser Lake area, and he started a survey of the Nechako River. Fraser Lake is connected to the Nechako River by the Nautley River, which flows out of Fraser Lake and joins the Nechako River in about one mile. The weather began to turn cold and snowy. At the end of October Swannell spent a night at the Hudson's Bay Company post at Fort Fraser. There he received a telegraph wire from the Surveyor-General that $3,000 had been sent through the bank so that he could pay his expenses. Swannell also noted that they had surveyed 373 miles of line. On November 1 there was a heavy snowstorm and the weather was very cold. As the inclement weather continued the men were forced to spend more time in camp. The river started to freeze and the men still had to get back to Quesnel. On November 6 Swannell completed the last of his river traverse. The next day he finished his township surveys and "got Polaris at elongation at MP. E Bdy 28 Tp. 13 Range 5. Make bearing of line N.0º03'W". His season in the Nechako Valley and Fraser Lake area was over.

By November 9 Swannell's crew arrived at Milne's Landing, a small settlement near present-day Vanderhoof. J.H. Gray's crew also arrived, and the men spent November 10 preparing for the trip down the Nechako and Fraser River to Quesnel. Swannell and Gray "divided the combined parties into 4 shifts of 6 men each," and left at 1 p.m. on November 11 on a scow that they had purchased.

> Run most of night – At 11 pm pass Grand Trunk Pacific crew camped 4 miles below mouth of Stuart River – Salute them with yells and rifle shots. Strike rock above Isle de Pierre Canon and stove in a plank in the bottom – Men below swarm on deck – manage to beach the scow on a gravel bar and land to repair– strip rind from bacon and nail over the shattered planking & have to put new valve in the pump. Desperately cold. River full of slush ice and skimming over in calm reaches. Put in at 2 am, but Smith, Benson, Wilkins, Swinerton, Sharkey, Mahood, Gray and I stay up all night.

The men started again at 6 a.m., reached Fort George at 5 p.m., and left Fort George at six the next morning. They passed through Fort George Canyon without difficulty, but:

> Very nearly smash up a few miles further down at Chila-Chula Rock. Can't control scow boat in the ice slush and scrape hard against the rocky islet – a projecting rock peeling off a long slice from our side. Excited yells from Capt. Dick "Hudson Bay boys, look out boys." River running ice very thickly. Camp 4 miles above Blackwater River. Smith, Wilkie and I sleep on board – when water leaking in reaches our feet, it wakes us up like an alarm clock.

The men reached Quesnel by late morning, stayed overnight, and proceeded down the Fraser River to Soda Creek the next day. Swannell made arrangements with the hotel manager to "look after and haul up scow which we leave in a backwater in front of the hotel…An American, Sackner, shoots up the hotel in approved American style – drunk and staggers off into the darkness. With Parson Allen round him up but decide to laugh it all off."

At Soda Creek Swannell and the surveyors switched to wagon transportation and traveled down the Cariboo Road to Ashcroft where they took the CPR to Vancouver. In Vancouver they boarded the *S.S. Charmer*, a CPR steamship, for Victoria, but a heavy wind forced the ship to lay up in the lee of James Island for several hours. They arrived in Victoria at 2:00 a.m. on November 21, ten days after leaving the Nechako Valley.

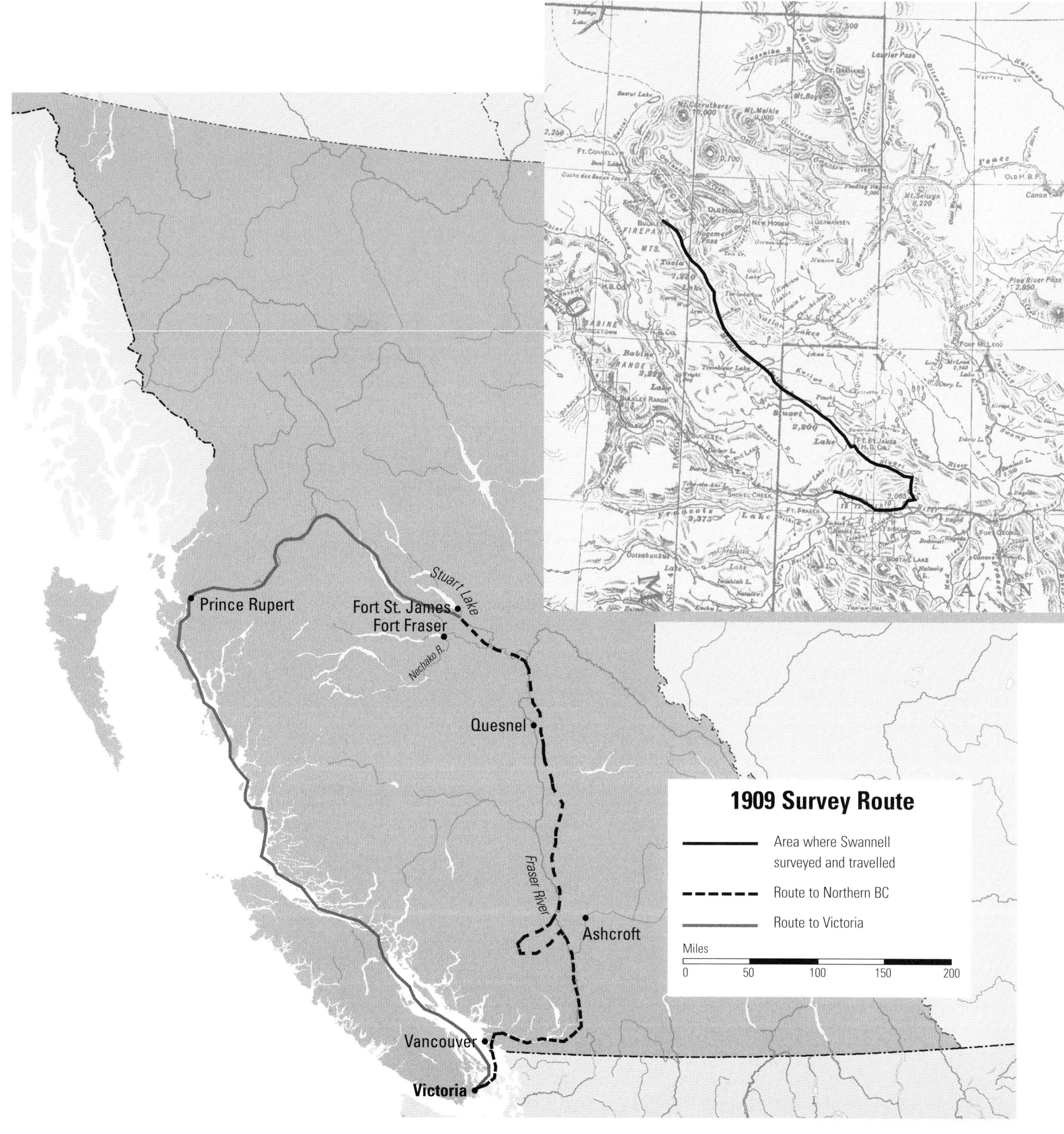

1909 Survey Route

— Area where Swannell surveyed and travelled
-- Route to Northern BC
— Route to Victoria

Miles
0 50 100 150 200

1909

"I am directed by the Hon., the Chief Commissioner of Lands, to instruct you to take up the unfinished work in the Nechako & Fraser Lake country." This opening sentence of the May 18 letter from E.B. McKay, the Surveyor-General for the British Columbia government, marked the start of another season in northern British Columbia for Frank Swannell. After providing some specific directions on the surveys to be conducted the Surveyor-General concluded, "I trust that you may have good weather and that other areas of land may be discovered during the season."

Two days later Swannell left Victoria for another field season of surveying which would last over five months and which would include over 2,000 miles of travel. This year A.I. Robertson remained in Victoria

■ Sliding down a snowslide on makeshift sleds. Schjelderup is in the front, while Swannell is behind him. Alf O'Meara took this photograph when the three men climbed a mountain during their Sunday leisure time.

■ Landing in Fort George Canyon. All the goods have been unloaded from the *Charlotte* and are waiting to be transported to the *Nechacco* at the head of the canyon. The horses are ready to pull a loaded sled across the skid trail. The sternwheelers that were built after the *Charlotte* were powerful enough to navigate Fort George Canyon so the necessity to transfer supplies soon ceased.

and Swannell was in charge of the surveying operations. One of his surveying assistants for the next three years was Vilhelm Schjelderup (BCLS #82). Schjelderup, who articled under Swannell, had a long, distinguished surveying career, most of it spent in the Nechako and Bulkley Valleys. Gerry Smedley Andrews commented that Schjelderup was known for the accuracy of his surveying. Another surveying assistant was George Vancouver Copley. Swannell and Copley had been close friends for several years. Copley worked as a surveying assistant for Swannell from 1909 to 1914. After 1914 he worked for the Forest Service, but remained a life-long friend of Swannell. In later years he and Swannell took Swannell's field notes for 1913 and 1914 and added many amusing and interesting details that provided invaluable insights into the daily life of a surveying crew and the work they performed.

Swannell, Schjelderup, Copley, and Alf O'Meara, a survey crew member who had previously worked with Swannell, "arrived Vancouver 9 a.m., 21st [May] having left Victoria on the midnight boat." That afternoon they took the train to Lytton, and the following day they trav-

■ *SS Nechacco* being repaired, Nechako River. Swannell's men are repairing some of the paddle blades that were damaged when the sternwheeler came through Isle Pierre Canyon. They were able to fix the *Nechacco* so that it could continue up the river.

eled by wagon to Lillooet. Swannell and his surveying crew spent the first part of the 1909 field season surveying lots in the Lillooet and Pemberton Meadows area. Meanwhile, R.M. Benson, who had worked for Swannell in 1908, went with two men to the Nechako Valley to begin surveying there. On June 3, Swannell went into Lillooet after completing some surveys. There he received a telegram. "Sharkey [Patrick Sharkey, the cook] wires horses split below Bobtail. Too impossible to catch. Wired Fraser [John A. Fraser, the Cariboo MLA who owned a grocery store], Quesnel – 'Our horses above lost. Arrange get my assistant Benson & 2 men into Nechako.'" Since Swannell was working on contract for the provincial government, he could request the assistance of the local MLA in finding replacements for the horses that had run away from Benson's crew while they were en route to the Nechako Valley. Sharkey probably sent the telegram from the Bobtail cabin, about 3/4 of the way along the Yukon Telegraph Trail from Quesnel to the Nechako Valley.

Throughout May and June Swannell's crew surveyed lots in several locations around Anderson and Seton

■ Ludwig's tent The *Nechacco* has stopped at one of the places where people have begun to settle on the Nechako River. Captain Bonser is in front of the tree with a jacket over his shoulder.

■ Sharkey in camp on the Nechako River. Patrick Sharkey was a cook on Swannell's survey crews for five years.

■ Hoy & Johnson's Cabin Swannell is on the left. Dave Hoy's partner, Johnson, is next to him. Dave Hoy is third from the left, followed by Bartlett, Dave's younger brother, Charles, and Anton Olsen, a settler. Dave Hoy spent six years in the Klondike working as a carpenter. He came to the Nechako Valley in 1906 where he owned a farm and cattle ranch. He drove cattle from the Chilcotin to his ranch where he would fatten the animals and then sell them to the Grand Trunk Pacific or local settlers. The *Cariboo Observer* reported on May 20, 1911: "Dave Hoy of *Nechako*, around here on Monday en route to the Chilcotin country after cattle, and proposes returning in a couple of weeks."

Lakes, and Pemberton Meadows. The men worked six days a week but still had the energy to spend some of their Sundays climbing mountains. On Sunday, May 30, "Schjelderup & I climb mountain about 6,500 [ft] – Leave 4 am – Snow deep on summit", Swannell wrote. His entry for Sunday, June 13, recorded that "Alf O'Meara, Schjelderup & I climb mountain ca [circa] 8,000'… - Glissade down slides. Fine view, glacier to south." Swannell recorded, on Sunday, July 4, that they climbed a mountain at the head of Owl Creek.

In contrast to 1908 in the Nechako Valley, Swannell did not encounter many difficulties with his surveys. There were only a couple of days of rain, and only a few diary entries mentioned mosquitoes or flies. The only serious mishap Swannell recorded occurred in the Pemberton valley. "Make a hay-wire raft tied together with rope & our braces – nearly get drowned under a log-jamb. When the bow hit the jamb the raft upended – Schjelderup jumped onto the jamb – I got my foot caught in the raft logs and went under the jamb - Schjelderup saw my wrist above the water, grabbed it & hauled me out – a very close shave." On this same survey Swannell's crew had to traverse a deep beaver pond. "Get across by driving axe into a big log to serve as a raft – very precarious transport."

By July 12 Swannell's crew had completed their surveys around Lillooet and they began their journey to the Nechako Valley. They traveled by carriage to Soda Creek and stayed overnight at the Soda Creek Hotel. They met the new proprietor and discovered that all of the goods that they had stored there the previous fall were gone. The next day the surveying crew departed for Quesnel on the sternwheeler *Charlotte*, as they had done the previous year.

At Quesnel Swannell found that the influx of settlers was already bringing changes to the region. Instead of a

■ Cleaning salmon at Stuart Lake. The dead salmon are lying on the shore of the lake.

Cleaning salmon at Stuart Lake.

ten-day horse trip along the Yukon Telegraph Trail, boat transportation was now available to the Nechako Valley. This also made it easier to obtain supplies for the surveyors. Swannell spent two days in Quesnel, and his arrival was noted in the July 24 edition of the *Cariboo Observer*. On Monday, July 19 he went to "Place order with HBCo & J.A. Fraser for supplies to be forwarded by steamer to Milnes Ldg - total 3,670 lbs. Finish Pemberton field notes & forward to Surveyor-General." The following day Swannell spent "Writing letters, arranging account with Northern Crown Bank – making an initial deposit $500". Then he left "Quesnel 4 pm sternwheeler 'Charlotte'. Reach Cottonwood Canon by evening and tie up for the night. OBN Wilkie [PLS #43] and party on board." The next morning they lined: "through Cottonwood Canon 6 am. Cable anchored at head of canon found in riverbed and cannot be picked up. Make thru China Rapids under steam. Arrive Canon Fort George in evening and find the "Nechacco" waiting at head of the canon. Transferring freight to "Nechacco" all day by stone-sled over skid road - mile long." The *Charlotte* was not powerful enough to travel through

■ Indians smoking salmon heads at Stuart Lake. This photograph shows the interior of one of the tents used by the First Nations people. Food is being prepared and the salmon heads are being smoked over the fire.

■ A Hudson Bay Boat at Fort St. James. This was the boat that the HBC used to transport goods and supplies on Stuart Lake. Items were taken up the lake to Portage where they were transported by wagon to Babine Lake. Another boat then took the goods to Fort Babine. The tramway that was built from the dock to the warehouse is visible.

Fort George Canyon so all the goods had to be unloaded and dragged by horses to the head of the canyon.

The sternwheeler *Nechacco* had been built in the spring of 1909 by the Fort George Navigation and Lumber Company to navigate the upper Fraser and Nechako River. It was piloted by Captain J.H. Bonser, a well-known veteran skipper who had handled sternwheelers on the Skeena and Yukon Rivers in Canada, and the Columbia River and its tributaries in the United States. The *Nechacco* made the trip up the Fraser and Nechako Rivers without difficulty, although the sternwheeler had to line its way through the Isle Pierre rapids on the Nechako River. Six days after leaving Quesnel Swannell's crew got off in the Nechako Valley at the cabin of David Hoy, one of the first settlers in the Nechako Valley. Swannell remained on the *Nechacco* while it proceeded up the river to Fraser Lake. The sternwheeler had to "line through three canons but the last two below Fraser Lake are too shallow. Unload freight. I walk up to Fort Fraser and have supper." There he met some of the settlers he knew from last year, including Jack Charleson, who had supplied beef for his men. After

conducting business at the Hudson's Bay post there, Swannell returned downstream on the *Nechacco* to rejoin his crew.

Swannell spent ten days in the Nechako Valley inspecting the work that Benson had completed and running some surveys himself. His journal entries for July 31 and August 7 mentioned frost. On August 9 Swannell and Copley departed for Fort St. James, the Hudson's Bay post on Stuart Lake, leaving Schjelderup in charge of completing the surveys in the Nechako Valley. Although Swannell does not provide any explanation in his field notes, he must have been satisfied with the progress and accuracy of the surveys and felt that Schjelderup could supervise the remaining work. The Surveyor-General's instructions had given Swannell the opportunity to explore for future surveying work in the region. Swannell had not been able to travel outside the Nechako Valley in 1908, and he probably welcomed the chance to see a new area and gain new adventures along with searching for possible surveys for the following field season.

Although it was only 35 miles overland from the Nechako Valley to Fort St. James, Swannell decided to travel by water. He rented a canoe from David Hoy, traveled 30 miles down the Nechako River to its confluence with the Stuart River, and headed 80 miles up the Stuart River to Stuart Lake. At the junction of the Nechako and Stuart Rivers Swannell and Copley visited Chinlac, a Carrier village that had been abandoned after the Chilcotins massacred many of the inhabitants in the village more than a century earlier. This massacre is

■ Landing freight at Fort St. James. The men in front are pulling a rope, while the men in the back are pushing the cart up the tramway.

Takla Lake 1909. Swannell took many photographs of his trip to Takla Lake. Copley is at the creek, and their canoe is on the beach.

Surveyors at the mouth of the Tache River. Swannell and Copley are on the left, while Devereux's surveying crew is on the right.

described in *The History of the Northern Interior of British Columbia*, written by Father A.G. Morice. They then had to pole and line their canoe up the Chinlac rapids. Above this they had to pole their canoe upstream for another day before the current became slack enough for them to paddle. Along the way Swannell and Copley met a Hudson's Bay canoe and four Carrier Natives "bringing down *quarterly* mail [four times a year] to Quesnel."

After eight days of upstream travel Swannell and Copley arrived at Fort St. James. They had supper with the Factor, A.C. Murray and his clerk, William Bunting, and slept in the schoolhouse at the fort. On August 20 Swannell wrote a letter to the Surveyor-General:

> I am on my way to Tremblai Lake in connection with some private surveying I expect to do in that region next season. While there I shall prospect Middle River & Takla Lake as I believe there is a large area of good land in that

HB boat, George Tod, an HBC employee, up mast at Fort St. James. This is a well-known Swannell picture that Fort St. James National Historic Park previously used on the cover of their park brochures.

Coccola and deer, Stuart River. Swannell is beside two deer that he has just shot. Father Coccola is going with the Fort George chief (in the white shirt) to Fort George. The other Native has a cross on his jacket.

vicinity which it would be advisable to place under reserve before it falls into the hands of speculators.

I am taking plenty of film along and trust to be able, on my return back, to give you a full report on the section in question illustrated by photos.

Fort St. James, at the foot of Stuart Lake, is the beginning of a vast waterway system. Stuart Lake is about 60 miles long. About 30 miles up the lake the Tache River leads up to Trembleur Lake, formerly called Tremblai or Tremblay Lake. Above this lake the Middle River connects with Takla Lake which is about 35 miles long. In the early 20th century this lake was more commonly called Tatla Lake. At the northwest end of Takla Lake the Driftwood River leads to the headwaters of the Stuart River system.

Swannell and Copley's trip through the Stuart River system was characteristic of their adventures during their six years together in northern British Columbia. The main purpose was to find new areas to survey since the provincial government had little information on the geography and resource potential of the region. At the same time, Swannell and Copley's quest for adventure and desire to travel took them far beyond the area they

originally intended to visit. Swannell used his camera to record the people and places that he visited.

Swannell and Copley traveled up Stuart Lake in one and a half days, then headed up the Tache River to Tremblai Lake. As they crossed the lake "very heavy swell in middle nearly split canoe." Tremblai Lake often has strong winds which create high waves that make the lake 'tremble.' Swannell noted Carrier homes at Grand Rapids on the Tache River and at the mouth of the Middle River. He and Copley found several fragments from the *Enterprise*, a steamer that had made its way up to Tremblai Lake in 1871 during the Omineca gold rush. (The *Enterprise* was intended to provide transportation for miners from the Fraser River up the Nechako and Stuart River systems to Takla Lake. From Takla Landing it was less than 50 miles to the gold fields in the Manson Creek area. On its first trip the *Enterprise* reached Takla Lake but broke up at Tremblai Lake on the return trip, and the boat was abandoned.)

After Tremblai Lake, Swannell and Copley canoed up the Middle River and traveled the entire length of Takla Lake. Swannell took many pictures of the scenery and caught fish on several days. Near West Landing, where the trail from Hazelton and Babine Lake to the Manson Creek gold fields crossed Takla Lake, he viewed a garden of "- acre rich loam - potatoes, cabbage, peas,

■ Sophie and Alice at Nautley Rancherie. This picture was taken at the Nautley Reserve near Fort Fraser. A cache is in the background.

turnips, lettuce, raspberries – all untouched by frost and looking fine." At the head of Takla Lake Swannell and Copley saw the remains of Bulkley House, a building constructed in the 1860s by the Collins Overland Telegraph. (Before the underwater Transatlantic Cable was completed, the Collins Overland Telegraph, also known as the Russian American Telegraph, hoped to establish telegraph communication to Europe from western United States through British Columbia and Alaska, across the Bering Strait and through Russia. Some members of the Collins Overland Telegraph spent the winter of 1865-66 at Bulkley House.) After meeting two mining prospectors, Swannell and Copley walked six miles up the trail along the Driftwood River.

Swannell and Copley spent eleven days on their trip up the Stuart River waterway. The return trip to Fort St.

■ Salmon weir at Fraser Lake. This elaborate weir was constructed across the Nautley River near the Reserve. Some of the channels in the middle where the salmon could pass through and be caught by the First Nations people are visible. In an effort to get the First Nations people to change from fishing to agriculture the Federal government banned these weirs in 1910. Copley and Swannell are standing on the weir.

■ Cataline's pack train, near Burns Lake cabin. Although Cataline was a packer for many years there are only a small number of pictures of his pack train and only a few of him.

James, starting on August 31, took six days. One of the Natives, Tremblai Lake Joe, steered them through the Grand Rapids on the Tache River. At the mouth of this river they met Frank Devereux's surveying crew. Devereux (LS) is considered the "grand old man of surveying" as he spent 71 years surveying BC's topography. Devereux, who had surveyed the Stoney Creek Reserve in the 1890s, was conducting several private timber and agricultural surveys in the upper Stuart Lake and Tremblai Lake areas. "Lunch at surveyor Frank Devereux's Camp at Tatche mouth. Cullerton, Mike, & Cook. Leave Mike & the Indian cook - bottle rum – and hear later that both were dead drunk and no supper when the line crews got back to camp."

Swannell and Copley reached Fort St. James on September 5 and stopped long enough to load 245 pounds of supplies before starting down the Stuart River. For most of the journey they had the company of Father Coccola, the famous Catholic Oblate priest, who was traveling with the Fort George chief and some other Carrier Natives. Father Coccola was based at Fort St. James, and his diocese stretched from Fort George to Hazelton, and north to Bear Lake, one of the headwaters of the Skeena River. Father Coccola continually traveled throughout the region, visiting the Native communities and settlers. In his diary entry for September 6 Swannell wrote, "Take easy day waiting for Father Coccola & Siwashes to catch up. They pass 6 pm and camp - mile below us. Coccola has the Fort George Chief with him". The Natives led Swannell and Copley through the Chinlac rapids, and Swannell noted that they covered the last eight miles to the Nechako River in 40 minutes. After departing from Father Coccola and the Natives, who were journeying down the Nechako River to Fort George, a day of upstream paddling brought Swannell and Copley back to David Hoy's place where ten people were crowded into the Hoy cabin for the night.

The next day September 9, Swannell visited Schjelderup's crew, and the following day he arrived at Benson's camp where he found "all hands in – epidemic of colds." Throughout the rest of September and early October Swannell and his crews finished their surveys in the Nechako Valley. The work proceded routinely and Swannell noted many days of fine weather. Just as the men began to leave the Nechako Valley on October 8, the first snowstorm of the autumn began. "Break camp – self, George, Rorison & Tom make to meadow 2 miles down the Stuart Lake trail – no tents & only one axe. Snowing most of day and all the night…we build a brush shelter but in spite of this snow drives in and we get out blankets wet. Rorison & Tom sleep in the snow under a pack cover." In his journal entry for the following day Swannell recorded, "Break camp early after a most miserable night – eight inches of wet snow on the trail."

Swannell chose to return to Victoria by a different

■ Yukon Telegraph Trail, Bulkley Valley. The trail and the telegraph line can be seen in this picture.

■ On the Skeena.

■ *SS Distributor* near Kitwanga, Skeena River. The *Distributor* was one of the sternwheelers that the Grand Trunk Pacific operated on the Skeena River between Hazelton and Prince Rupert until the completion of the railroad through this area.

route from the previous season. The men traveled by horse along the Yukon Telegraph Trail to Hazelton. Swannell's diary for the trip is missing, but he noted that on October 23 they were at Glacier House and on October 24 they were "At Morice Town - stay in Indian Bunkhouse." Swannell photographed sections of the Yukon Telegraph Trail and the pack train of the well-known packer Cataline. At Hazelton Swannell and his survey crew boarded a sternwheeler for Prince Rupert. Swannell took pictures of the new and growing town that was being built at the Pacific terminus of the Grand Trunk Railway. From Prince Rupert the men traveled by steamer to Victoria, returning during the first week of November.

■ Waterfront, Prince Rupert. In 1909 Swannell took several photographs of the new city of Prince Rupert that was the Pacific terminus for the Grand Trunk railroad.

■ Prince Rupert 1909. This picture shows a street in Prince Rupert with a mixture of temporary dwellings and new homes.

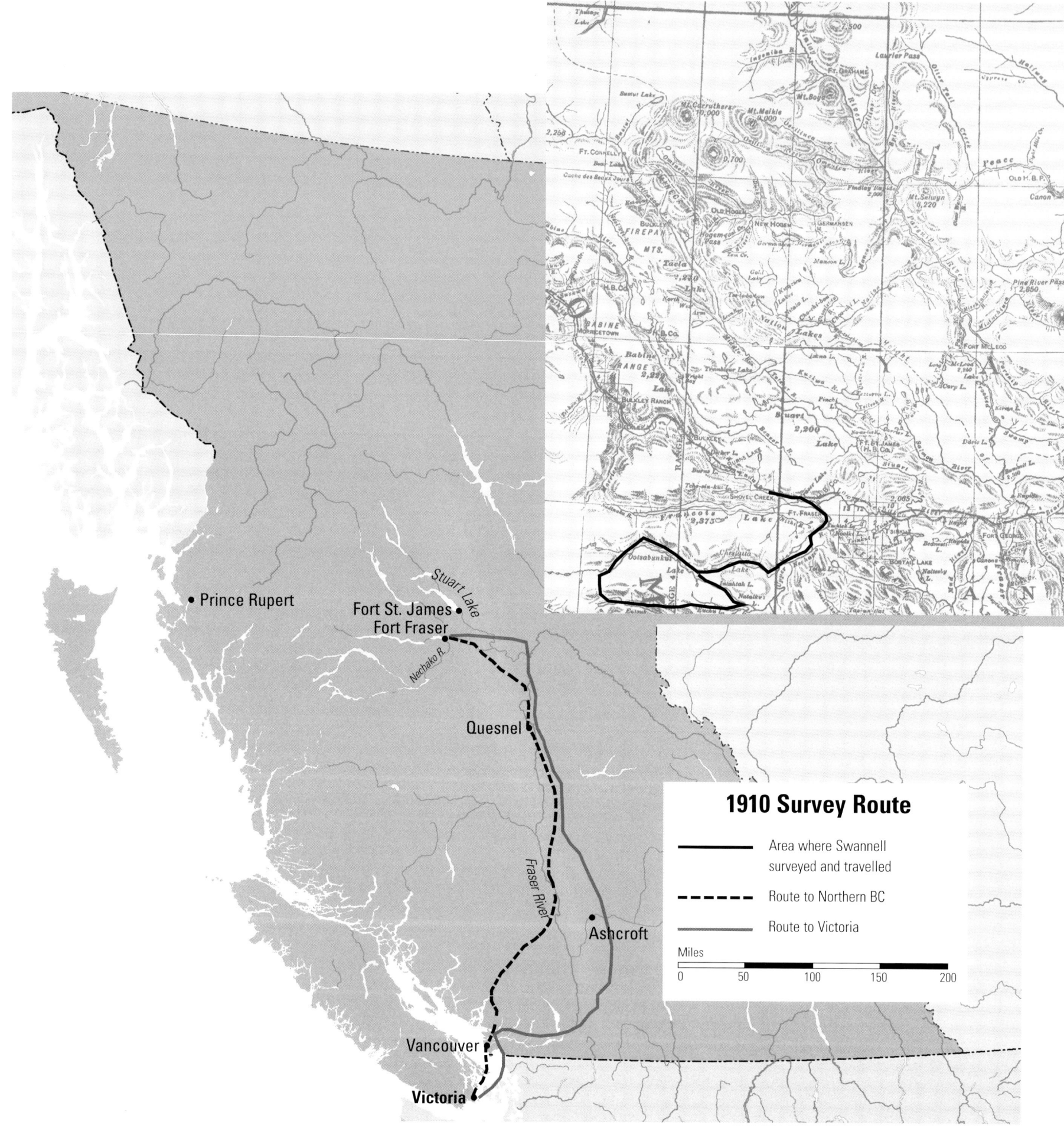

1910

In 1910 the provincial economy continued to prosper, and the demand for land led to a need for many surveys in British Columbia. A.I. Robertson started his own surveying business, so Swannell went into partnership with A.O. Noakes (BCLS #39) who received his surveying licence early that year. Noakes managed their office in Victoria, did the drafting and calculations for the surveys, and made some surveys around Victoria.

During 1910 Swannell was engaged in surveys in several locations around British Columbia. In January and February Swannell and Schjelderup worked around Victoria. One survey took them to Saltspring Island. "Complete survey and leave 10:30 p.m. on Bittancourt's launch. Very windy, and have to anchor in lee of Sidney Island, and reach Victoria 9 a.m." At the end of January Swannell traveled to Alberni for a survey. "Leave Nanaimo 8 a.m. in Studebaker automobile – rain & snow all day – at summit smash axle in 16" snow – Drive rest of way in rig from Alberni – driver drunk." On the return trip, Swannell wrote that, "At summit stopped by blast [of] 12 holes, throws about 40 tons of rock and three big fir across the road – Swede station-men clear the road and carry the rig across the debris."

In late February and early March Swannell and Schjelderup, along with a surveying crew and some forestry cruisers, went to Gordon River on a trip that was filled with misadventures. On February 21 Schjelderup and the cruisers traveled from Victoria to Port Renfrew

■ Fording the Upper Lillooet River. This picture and the following one show some of the difficult conditions Swannell and his crew encountered while surveying in the Pemberton and upper Lillooet River area in the early spring.

on a tug. Swannell and two men from his surveying crew boarded the *Amur* at 11 p.m. and tried to sleep while freight was being loaded all night. After a four-hour trip the next morning Swannell and one man took a canoe from Port Renfrew to the mouth of the Gordon River while the cruisers and surveyors packed their supplies to the base camp. In canoeing up the Gordon River Swannell noted, "Have to jump overboard and haul the canoe up the riffles – water icy." The next day they found that all of their sugar had gotten wet and spoiled, and in a steady rain they made a cache and moved their supplies. On February 25 Swannell and three men attempted to cross the Gordon River on a cable basket. "The platform at the end of the cable breaks. Jack [Young] and I jump to the bank. The cruiser falls 15 ft and the compassman hangs on to the cable." With heavy snow and sleet the river rose three feet the following day, and four men quit. "Tent leaking badly, spoils blue-prints. Rig up fly and am happy." On the 27th and 28th of February heavy snow and rain continued. The river rose another three feet in four hours and the men were forced to move to a cabin across the river for fear of a freshet. After two more days of heavy rain and snow, Swannell decided to abandon the survey on March 2. The men packed out on March 3 in pouring rain with very little surveying completed.

Camp at Big Point, Upper Lillooet.

In mid-March "Tom Greer, of Pemberton arrives to lead us over the Coast Range to the meadows. At supper, being asked to have a second helping makes the famous reply, "Mrs. Swannell, I have had an ample sufficiency; any more would be a superplenitude of abundance". On March 16 Swannell, Schjelderup and the surveying crew departed for Vancouver where they spent the day outfitting. The next morning the men left Vancouver on the "SS Britannia, stopping at Anvil Island & Britannia Mine. Arrive at Newport 2 pm and go by stage 8 miles to Brackendale." The *SS Britannia* was the flagship of the Terminal Steamship line. The following day they had to put on their snowshoes. On the 19th Swannell wrote that there was "five feet of snow, but creeks open and very difficult to cross." The next day they reached Summit Lake, also known as Alta Lake, in the Whistler area, where their surveying started. By the 24th of March the crew had "No grub except flour & beans. I fall through 6 feet of snow into the creek snowshoes and all and couldn't get out; but managed to attract attention by waving my axe handle; being completely out of sight myself." Sunday dinner on the 27th was spent at Bauers in the Pemberton Valley. "Bauer is an old sergeant U.S. cavalry out west…Joe yarns of the Indian days." The next day the surveys took Swannell and his crew up the Lillooet River where they found "mosquitoes very thick – in deep snow & March!"

During the field season the surveyors worked long, hard hours six days a week. Sunday was a time for rest and relaxation, and sometimes an opportunity to visit with some of the settlers in the area where they were working. April 3 was spent:

> At Ronaynes all day copying notes and writing letters. Important trial before John Ronayne J.P. At Alta Lake we had taken 5 lbs. of rice from a cabin as we were out of grub The owner of the cabin, hearing of this, had a Vancouver lawyer write John Ronayne asking him to take legal action – all the Pemberton settlers thought this a huge joke. Special constables were appointed who dragged in Tom with a bag of rice hung around his neck. Tom was sentenced to imprisonment in the cellar but released as he was caught making inroads on Mrs. Ronayne's preserves.

On the 5th of April they were surveying lots in the upper Lillooet River area. "Ford the Upper Lillooet 3 times. Water 2 - ft. deep, and nearly swept away. Snow at banks 5' sheer. Rain part of day and very heavy snowshoeing." By the 8th there was no food remaining except

■ Frank Swannell with two pair of snowshoes. This picture shows the two pair of snowshoes that Swannell used during his surveying. The larger pair of snowshoes were used for traveling, while the bearpaws, the smaller snowshoes, were used when Swannell was running the transit.

for potatoes and flour. Swannell noted that in "all these jobs it has been desperately cold standing behind the transit in soaking wet moccasins". Despite the difficult conditions Swannell, Schjelderup and the rest of the crew steadily continued surveying lots throughout the Pemberton area in April. By April 22 the surveyors had completed their work in the area, and they traveled to Anderson Lake where they sailed down the lake. They stopped overnight at the Rancherie (Native Reserve) where they had a supper of salmon, rice and bannock. The following day the men rowed the full length of Seton Lake and arrived at Lillooet at noon.

Swannell left Schjelderup and the rest of the crew to survey some lots in the Bridge River area while he

■ Group on Seton Lake, and the steamer that operated on Seton Lake. Swannell's wife, Ada, is in the back near the smokestack.

returned to Vancouver. On Sunday, April 24, "Tyee Jimmy drives me through to Lytton, $10.00" but on Monday, "Held in Lytton until 3 p.m. – train held up owing to a washout 3 miles W. of Lytton – Passengers transferred across gap. Arrive Vancouver midnight." The next day his wife, Ada, arrived on the morning boat and he interviewed Mackay, Wright & McIntosh from the Nechako Settlers Association regarding some surveys in the Nechako Valley before returning to Victoria.

There he met with Price Ellison, Minister of Lands, and McKay, the Surveyor-General regarding surveying contracts with the provincial government for the 1910 field season. Swannell must have received assurance that he would receive the contracts, for on May 9 Trygve Rognaas left for Quesnel to prepare for the Nechako Road contract. This contract was for the construction of a wagon road from Fort George to the Nechako Valley. Rognaas worked for Swannell during the 1910 and 1911 seasons. He made a map of the upper Nechako River area for the provincial government based on Swannell's surveys. Rognaas was the cartographer for T.H. Taylor's coal surveys in the Groundhog region of northwestern BC from 1912 to 1914. On May 9 the Surveyor-General requested an additional survey. "I beg to enclose tracing of a small reserve in the Pemberton Portage between Seaton and Anderson Lakes, and as you have a survey party in the Pemberton Valley and are going to join them, I wish you to subdivide the land shown on this reserve into 80 acre blocks, arranging the lots to the best advantage, with respect to the existing Indian Reserve

■ Freight team, Cariboo Road. A classic picture of an eight horse freight wagon operating on the Cariboo Wagon Road. This photograph was taken at the end of this era of transportation. Within a few years automobiles and trucks became the main mode of travel.

■ Freight wagons and automobiles at 153 Mile House. Automobiles were beginning to appear on the Cariboo Wagon Road, and their presence often spooked the horses. This freight wagon has gone off the road and is damaged. The 2 on one of the automobiles indicates that these vehicles were part of the commercial automobile service that was just beginning in the Cariboo in 1910.

■ Central Fort George. This photograph shows the beginning of construction in a community that would become part of Prince George in a few years. The cutbank along the river is an identifiable landmark.

and any land which may be held by pre-emptors." On May 11, Copley left for Lillooet to prepare for this survey.

The same day the Minister of Lands met with Swannell. "Am hauled over the coals by Price Ellison & Renwick, Deputy Min. re the Leduke AP [application to purchase]. Find out the latters P.M. [pre-emption] has been sold over his head." Swannell did not provide details, but the Minister was evidently unhappy with his handling of this survey. The provincial government's displeasure was also expressed in the letter that the Surveyor-General sent to him that day. Although he confirmed Swannell's contracts, E.B. McKay's letter lacked the friendliness of the previous year and contained some direct warnings. "You will be required to keep within this Reserve [Upper Nechako] as no surveys outside of such reserve will be accepted or paid for by this Department." Probably referring to the previous year, where Swannell had left much of the surveying to Schjelderup while he and Copley traveled to the Stuart Lake country, McKay wrote, "It is to be distinctly understood that you personally superintend the whole of this work, and no information as to the nature or extent of land is to be given out except to this Department." At the close of his letter McKay warned Swannell, "I have now only to wish you a successful season and trust that the nature and extent of the lands surveyed by you will justify His Honour the Commissioner in making your appointment."

Swannell now had to organize supplies and men for the various surveys with which he was involved. Rognaas needed a crew for the Nechako Road survey; arrangements had to be made for surveying a Native Reserve at Seton Lake; and preparations for another field season in the Nechako Valley had to be started. On May 16 Swannell hired men and purchased equipment for

Blair's Store, Central Fort George. William Blair was a prominent businessman who owned stores at Fort George, in the Nechako Valley, and in the Cariboo.

Rognaas' road survey. Three days later, W.E. Waters, an English civil engineer, left with a crew of men to join Copley and begin the Seton Lake Reserve survey. On May 21 Swannell "saw Manager Northern Crown Bank and made arrangements for a possible overdraft of $3500."

In late May Swannell left Victoria and traveled to Lillooet. On May 25 he "went up to Short Portage by steamer, lunching at Waters Camp. Deputation of Indians interviewed me with view to proving all Short Portage Indian land. Had paper from Indian Agent Measton to that effect." After meeting with the First Nations people and examining the area, Swannell decided not to continue surveying the Reserve. On the 4th of June he wrote to Archibald MacDonald, the MLA for Lillooet:.

I enclose herewith sketch shewing surveying done to date on the Short Portage in pursuance of instructions given me by the Surveyor-General. We met with considerable opposition from the Indians, who claim that the whole Portage is theirs. The other day they shewed me a document, of which I enclose copy. Of course the Indian agent had no right to give them such a statement, as he had no authority to declare a reserve. Nevertheless the Indians doubtless think it is all right. They have gardens and

■ *SS Chilco* at Giscome Portage on the Fraser River. The supplies have been unloaded and are waiting to be taken up to the Huble farmstead at the top of the hill where Swannell took his photograph. The cleared area by the sternwheeler is a vegetable garden.

buildings scattered all over the area between the two surveyed reserves, and I think should at any rate be compensated if these lands are to be taken away from them…I have to leave for Victoria tomorrow and so am not taking the survey any farther until I have consulted the Surveyor-General & heard your opinion on the matter. As you will observe from the sketch much of the reserve is mountain – In fact I do not think we could get more than 3 or 4 eighty acre lots without running into useless mountain side.

During the next two weeks in June Swannell continued making arrangements for his surveying crews. The B.C. government decided to proceed with surveying the Reserve at Short Portage, for on the 19th Swannell left Victoria with his wife, Ada, for Lillooet. This was one of the few occasions that she spent some time with her husband when he was surveying in the field. From Lillooet they traveled down Seton Lake to Short Portage where they met Schjelderup. In a week the surveyors completed the Reserve along with several other lots in the Lillooet area. One of these was for Charlie Adolph at Fountain. In January, F. Souses, the Government Agent at Clinton, had written to Swannell telling him of some surveys that needed to be done in his district. One of these was for Charlie Adolph. The previous surveyor had

■ The *SS Chilco* at Isle de Pierre Canyon, Nechako River.

SS Chilco, Upper Nechako River. Swannell's crew is getting ready to get off the *Chilco* after its famous trip, the only time a sternwheeler was taken on the Nechako River above Fort Fraser. Captain Bonser is on the far left of the third level with his wife beside him. The cordwood is stacked on the bottom level, with a cross-cut saw at the front.

charged Charlie for the survey but had never registered it. The Government Agent warned Swannell that Charlie Adolph is "a good Indian and you deal gently with him in the matter of charging him for the survey, or I may have something to say to you when you arrive here."

On July 1 the surveyors left Lillooet. The trip north was made in a new form of transportation that was beginning to appear in the Interior of British Columbia – the automobile. Two automobiles were hired for the twelve men. On the first day they traveled to 141 Mile House, stopping en route at Clinton where Swannell met with the Government Agent and MLA. The only inconvenience was the necessity to put on chains on the 100 Mile hill. The next morning they traveled to Soda Creek where the *Quesnel* took them to Quesnel. This sternwheeler had been constructed at Quesnel in 1909. Along the river they passed the *BX*, another new boat, and the *Charlotte*.

At Quesnel the surveying group split, with Schjelderup taking most of the men to Fort George, while Swannell and a few men surveyed some lots around Quesnel. Swannell finished the plan and notes for the Indian Reserve at Short Portage and also made his summer arrangements. "Interviewed Cameron, Mgr. Northern Crown Bank and gave him cheques $1500. Transfer Victoria a/c $1000 to be sent now, balance on word from me." The field notes for some of the lots surveyed were sent to Noakes, while 1800 pounds of supplies were ordered, to be sent to the Hudson's Bay post at Fort Fraser, and 1000 pounds were shipped to Rognaas.

On July 19 all of the survey crew, except Swannell and Copley, left by wagon with supplies for Schjelderup's crew, which was now working along the Nechako River. The next evening Swannell and Copley boarded the *BX* for Fort George. The *BX*, which had been launched that spring, had been built by the BC Express Company, which provided stagecoach service in the Cariboo. It was the largest, most powerful, and most luxuriant sternwheeler on the upper Fraser River. "Along the Fraser River "Passed the wreck of the *Charlotte* today at Fort George Canyon – Pretty badly smashed pounding in rough water, lying on a bank, stern submerged. Cable had snapped near capstan." At Fort George Swannell saw the beginnings of one of the main centers along the Grand Trunk Pacific in British Columbia at the confluence of the Nechako and Fraser Rivers. In a few years this location would develop into the city of Prince George.

On the 26th Swannell left on the *Fort Fraser*. This sternwheeler, constructed in 1910 by the Fort George Navigation and Lumber Company, was known as the prospector's boat and was the smallest sternwheeler on the upper Fraser and Nechako Rivers. Ten miles up the Nechako River the engine broke down and the *Fort Fraser* returned to Fort George. Swannell was anxious to visit Rognaas and inspect his work so the next morning "Self & Bjornfelt, a young Swede, leave by trail up the Nechako for Rognaas Road Location camp at 5:30 a.m. Breakfast at Slim Millers, 14 miles, Mud River – Bjornfelt plays out 23 miles out. But I leave him and push on, and finally, altho' desperately tired reach Rognaas' Camp at Cluculz – a record over this trail as over 40 miles and I had a light pack for the first 23 miles."

After spending several days with Rognaas, Swannell returned to Fort George. In early August Swannell took a boat trip on the *Chilco* (the *Nechacco* had been renamed in 1910) with Captain Bonser up the Fraser River to the Huble farmstead at Giscome Portage. From Giscombe Portage, on the Pacific watershed, there was a

■ Wooding up, Upper Nechako River, Swannell's crew are cutting trees for firewood for the *Chilco*.

Grand Canyon of the Upper Nechako River. Swannell took several photographs of this section of the river. Kenney Dam was constructed at the top end of the canyon, and the Skins Spillway sends water into the Nechako River below the canyon so this well-known area of the river is now mainly dry.

nine-mile trail to Summit Lake at the head of the Crooked River, on the Arctic watershed.

On August 12 Swannell and Copley departed Fort George. Copley, who wrote a magazine article for *B.C. Outdoors* about the 1910 field season, described how Captain Bonser took the *Chilco* up through Isle Pierre Canyon.

All hands off to drag a one-inch steel cable up to the head of the rapids to be made fast to a sizable spruce tree. The mate went with us to make sure that our hitch on the tree was secure, then the wind-up commenced. All went well until the steamer was about one-half way up the rapids, then without warning the spruce tree came out, roots and all. The steamer slewed around crosswise to the stream and down we went bumpity bump over the rocks. Water poured over the lower deck and into the fire box, so that when we finally arrived at the foot of the rapids into good water, the ship had no power. Luck was with us, however, and just as the anchor was thrown overboard we landed bow first on a sandbar. The fireman and engineer soon got a fire started again and in less than two hours we were back at the foot of the rapids, the captain as unperturbed as though the whole matter had been routine. The next try we fastened to a larger tree and had no further difficulty getting up the rapids.

Besides piloting the sternwheeler Captain Bonser helped procure the food for his passengers.

Another day, while steaming along in fair water, the captain saw a nice deer standing on a rock cliff staring down at the boat. He simply let the wheel go, grabbed his rifle, which he always kept in the wheelhouse, shot the deer and then had the ship straightened out before it had time to become crosswise in the stream. He then edged the steamer to the river bank, tied up to a tree, and we had fresh liver and venison steaks for supper.

After a two-day trip up the Nechako River, Swannell arrived at Milne's Landing. The next day they visited Hoy and Johnson, whom they had met in 1909, and the following morning "Copley & I leave 10 a.m. – river very high and difficult poling. At Noonla, when lining, the bridle on the line snaps, the canoe slews and dips under – one sack of sugar spoiled and box containing spare transit and medical supplies lost." They proceeded up to Fort Fraser, where they left the supplies they had brought, and then visited Schjelderup's camp. On August 19, when he was back at Fort Fraser, Swannell recorded that he had received "telegram 'steamer will arrive Sunday afternoon.' Arrange with trusty Indian retainer Wm Ketlo to take Ross' horse and ride to Schjelderup with the message." On Sunday "Schjelderup's crew arrive from Stella – some by trail with horses, others in boat of Stella chief Isidore. Pitch camp at the Nechako ferry." This was Vital's ferry on the Nechako River, near Fort Fraser, where the *Chilco* would land.

On Tuesday, August 23 Captain Bonser and the

■ Chief Louis & family, Ootsa Lake. Chief Louis is on the left, with his family at the east end of Ootsa Lake by the portage from Cheslatta Lake.

■ Whitesail Lake. This is the canoe of Cheslatta Edmond, with whom Swannell exchanged provisions. The family is traveling in a dugout canoe. Hudson's Bay blankets are in the middle.

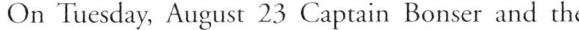

Chilco arrived. According to Copley, Swannell, Bonser, and "another man who knew the upper river, got together over a bottle of rum to discuss the idea of taking the steamer on up another 50 miles. It was finally decided that if we would cut the necessary wood to steam up the river and return, the captain would take his boat up the river as far as he could without endangering the craft." Probably Bonser wanted to enhance his reputation, for he was a daring river pilot who claimed that he could take his boat anywhere there was enough water to keep the hull from scraping bottom. In his journal Swannell wrote:

> "SS "Chilco", Capt. Bonser takes my survey party up the Nechako above the Fraser Lake Ferry to the foot of the 1st canon. We are 20 in all – and this is the first and only time a sternwheeler was ever taken up these waters. Turned the crew [of] mine loose ashore to wood up & Capt. Bonser is much impressed with the activities of my men in downing trees – so delighted in fact that he only charges me $75 for the trip.

The trip went without incident, and in a day and a half the survey crew got off near the foot of the Grand Canyon of the Nechako, not far from the present-day Kenney Dam.

Twenty men in a surveying camp consume large quantities of food. Despite all the supplies purchased in Quesnel, Swannell soon found that he would need more food. After a week of surveying Swannell and Tom Greer canoed down the Nechako to Fort Fraser where they bought over 600 pounds of meat and vegetables. "Wired Surveyor-General: 'To date surveyed 10,000 acres & 10 miles river traverse. Request $1200 placed my credit Northern Crown Victoria. Wire confirmation." He

■ Lagoon on Whitesail Lake. Swannell took many photographs of the lakes and waterways on which he traveled during the Great Circle trip. Since this area has been altered by Kenney Dam these pictures are an invaluable record of the geography of the upper Nechako region in the early 20th century.

"wrote J.A. Fraser (MLA) Quesnel to get for me $3000 for Rognaas Road Survey downriver. Swannell needed these funds from the provincial government to pay for the expenses of his surveying crew. On September 2 "Tom Greer, Anton Olson (Norse land staker) & I leave Fraser Lake in poling boat 9 am with 270 lbs beef & 400 fresh vegetables. Make to surveyor J.H. Gray's camp at 6 o'clock – about 15 miles. Boat is overloaded, only 3" gunwale and poled heavily." The canoe was carrying so many supplies it was just barely floating above water, and it could have easily been swamped by a large wave.

By the 4th they had arrived at camp. "Send wire down by Indian Patrick advising Noakes that prospecting will commence at once. Schjelderup and Copley chain Gray's line and find it a chain in error." The next day the main camp was moved upstream. "Olson and I left camp in Chestnut canoe loaded to 2" of gunwale, tried to cross river in eddy above 2nd canon, swamped and went through the canon sitting up to upper vest pocket in water. Made landing below the canon – one tin of potatoes floated off – everything soaked…spent evening drying out our blankets, tobacco and saveable grub."

Although the Surveyor-General had specifically stat-

Swannell's survey crew, upper Nechako River. The men are having fun posing for this picture.

ed that Swannell was to personally superintend all of the surveying, Swannell now split his crew. Schjelderup took the majority of the men to continue on the surveys along the upper Nechako River while Swannell and five men left for Cheslatta Lake on September 6 to make a triangulation survey of the lake. On the way up the Nechako River the "boat got away when lining the little canon below Cheslatta Ck. – and ran thru the canon holding a few minutes in an eddy on the other side – Olson swims out for it when it gets into an eddy our side." Before going up to Cheslatta Lake, Swannell visited and photographed the Grand Canyon of the Nechako, a section of the river that was altered when Kenney Dam was constructed in the 1950s.

The triangulation survey of Cheslatta Lake was very similar to Swannell's survey of Fraser Lake in 1908. There was a Native Reserve at Cheslatta Lake and a new one was going to be established, so the provincial government probably wanted detail of the size and shape of the lake along with the surrounding lakes, creeks and mountains. On September 9, their first day at Cheslatta Lake, it rained most of the day. "Holiday in camp. Olson & I climb a mountain and get a good view of a low valley behind the Rancherie – meadows & small lake visible." The next day Swannell started the triangulation and took an observation of Polaris for direction. The following day they "measured east base and took observation." On a typical day Swannell and Olson would be reading the angles while the other three men would be setting signals at points along the lake or on nearby hills. The triangulation of Cheslatta Lake proceeded smoothly and was completed in about two weeks.

Then Swannell, Tom Greer and Anton Olson left on the Great Circle trip through the lake country of the upper Nechako Valley. Today this area is the Nechako Reservoir, held back by Kenney Dam. There are two main valleys, each over 50 miles long. At the western end the two valleys almost join each other, separated by an easy portage that is less than one mile long. In 1910 there were several lakes and connecting waterways in each valley.

On Friday, September 23, while "Dan & self chain & read West end base" to complete the triangulation, the other men took "the poling boat across to Ootsa Lake by horse & travois" before departing. On their first night at Ootsa Lake they camped at "Louis' Dumping Ground at end of trail." While traveling on Ootsa Lake they passed the surveying outfit of E.P. Colley (PLS #67). Colley surveyed much of the land in the Francois Lake region and the lakes of the Great Circle area. While returning from his annual trip to Britain he died in 1912, one of the people who perished on the Titanic. Along Whitesail Lake they met "Cheslatta Edmond and exchange sugar & tea for dried goat meat. We find Father Morice's portage and get canoe across into a little beaver lake. 1st portage 350 yds., second 1/3 mile into St. Thomas Bay, Eutsuk Lake. Raining all afternoon." Now they were on the return part of the circle, and with

■ Mr. and Mrs. William Bunting. This is one of two photographs that Swannell took of the Buntings when he visited the Hudson's Bay Company store at Fort Fraser at the end of October. The Buntings are already dressed in winter clothing. The fur press is visible behind Mr. Bunting. William Bunting came from Scotland in 1907 to work for the HBC. Originally he was posted to Quesnel. In 1909 he was transferred to Fort St. James, and in the fall of 1909 he was promoted to Factor of the Fort Fraser post. After Fort Fraser closed in 1915 Bunting returned to Fort St. James. In 1920 he opened a store in Fort Fraser that operated for many years.

a favourable wind they were able to sail down much of Eutsuk Lake. Swannell noted that the land along Tetachuk, the next lake, was completed staked on both sides. At the outlet of the lake the Tetachuk River was filled with rapids.

> We ran the first rapid, struck a boulder, broached and partly swamped -jumped overboard to clear and managed to get down an unexpected cascade – a 5' drop – safely. Then Olson & Tom lined and poled down light, making one portage at the lower cascade, 20 ft. sheer. 4 cascades above within a mile with rapids between – cascades average 10-25' fall. Greer grub box was lost when the boat swamped dropping down. Noticed old canoe wedged crosswise in rocks at head of the upper cascade... Buckwheat plays out – only a little rice, goat & venison, 2 inches of bacon & 1/4 ham left

Swannell noted the Bella Coola trail that crossed near the foot of the Tetachuk River. The Carrier of the Nechako area used this trail for trading and contact with the First Nations tribes of the Pacific coast. Further south the Bella Coola trail joined the Grease Trail that Alexander Mackenzie used on his route to the Pacific. By October 10 Swannell, Greer, and Olson were back to Ootsa Lake where they met "Colley's outfit heading for Bella Coola and get 25 lbs flour and some rice and prunes." The next day they arranged "with Baptiste

■ Surveyors on lower canyon of upper Nechako River. Swannell and his survey crew are getting ready to head down the Nechako River on their way to the surveys at Endako.

Frank Swannell at transit in snow. Swannell is using two pair of mittens, a thin pair for working with the transit and writing in the field book, and a second pair that he could put over the first when he wasn't working. This photograph was taken in December in the Endako Valley.

'Broken Nose' ,Chief Louis' son, [his brother] to take us across to Cheslatta Lake, consideration being my binoculars ('glass eye') & $2 cash." It took Baptiste a day to round up the horses. Back on Cheslatta Lake they "bought at Rancherie white-fish & a 'nun-ti' (arrowhead)." By October 15 Swannell, Greer and Olson returned to the survey camp on the upper Nechako River.

It was now mid-October and getting late in the season for surveying. Swannell divided the men into three parties with Schjelderup, W.E. Waters, and Copley each heading one group. On the 19th of October they moved camp down the Nechako River.

Charlie & Roy slew round in poling at a gravel island, are swept under bushes and canoe capsized. Everything dumped into river – quarter beef, plates, pots, cups, case of fruit & one axe lost. Some of the plates & cups and the big tent salvaged. Supper late as there are no pots nor grub to speak of. When we arrived in camp Roy & Charlie were cutting rounds off a felled poplar for plates. My first query, hearing of the accident was 'Anybody drowned!' Roy's reply 'Far worse than that, cook outfit is lost.'

On Sunday, their day of rest, "Take photos of crew in big tent and later of all in canoes on the river. During the excitement Olaf & Pike in the small canoe capsized – but I was not quick enough to get a snapshot so they very obligingly stage it all over. Very sporting as the river is icy cold."

At the end of October Schjelderup and Waters and their crews left by canoe for Vancouver, while Swannell, Copley and one crew remained. Swannell and two men went down to Fraser Lake. On the first day it was "snowing hard most of day and lose our way cutting across country 7 miles to west end Hallett Lake – have to steer our way by compass…camp at dark. Keep a fire going but spend a miserable night under a spruce." The next day they arrived at Fraser Lake in the late afternoon. "Pay Siwash 50 cts for crossing us at the Fish Trap. Put up at Buntings [The Hudson's Bay Company factor]. The other two men left for Stoney Creek while Swannell spent two days at the Fort taking care of business matters. On November 3:

I leave the Fort 6 a.m., snowed heavily nearly all morning. Very heavy walking with 10 inches snow on the trail summit. Reached Hallett Creek 7:15 p.m. very nearly played out; it being very hard to keep the trail in the dark & snow. Started to hunt for a dry stump and firewood in the dark. Heard a shout and Olaf Larson turned up – He had come out to meet me with a little food *but* no blankets. A splendid fellow. Spent a fair night by keeping a fire going.

The next day Swannell and Larson returned to camp. At their camp Copley took a cross-section of the Nechako River and estimated that the average flow was 11,000 cubic feet per second, a big contrast to the average water flows after Kenney Dam was constructed, which are normally under 5,000 cubic feet. Even in the upper part the Nechako was an impressive river.

During November Swannell's crew made triangulation surveys around Hallett, Copley and Rognaas Lakes, and surveyed several lots in the area. Swannell noted cold weather and snow in several of his journal entries. November 9: "Snowing part of day and very cold…Spent miserable day at the transit with only one mitt and shoe packs frozen solid. Charlie (cook) makes a stove for the sleeping tent out of a bake pan & reflector. Lake froze nearly all over last night." The following day "Copley & I also climbed mt. and re-erect signal. Bitterly cold on top. Snowed two inches in night and heavy snowstorm morning." On the 11th Swannell wrote, "Lake completely frozen over, half being glare ice." By the middle of November the transit was "continually freezing up and have to build fires to thaw it out." On the 21st Swannell noted that there was a foot of snow on the ground. Despite these conditions their survey lines were accurate.

In his notes for November 25 Swannell wrote, "Very cold night and morning – dare not touch the transit with

■ Survey party on the Nechako Road using Constantineau's sleigh. Some of the survey crew are getting ready to travel on Constantineau's sleigh. The coats are made from Hudson's Bay blankets.

bare hands. River starts running ice. We steer our course to work by moonlight…Impossible to leave for work before 7:45 a.m." The next day was a "bitterly cold day. Olaf gets his feet frostbitten. Build fire at eleven to thaw ourselves out. Poor Olaf has a hard time – his oil-tan shoe packs freezing into sheet-iron. He stamps around howling like a wolf…River running ice all day and freezing over in back eddies." Swannell realized that they were not going to complete their surveys on the upper Nechako River this year. On November 28 the surveyors made preparation for going down the river. A cache was built and about 1200 pounds of food and some equipment were stored in preparation for an anticipated return to the area next year. In a note added later to his journal Swannell wrote, "This cache was found as we had expected untouched next year. But a Fraser Lake Indian came to me in 1911 to confess that starving he had broken in during the winter and taken out 5 lbs of rice. He felt very guilty."

The trip down the upper Nechako River went without incident but ice in the river made navigation difficult. In one section of the river the boat got "locked in a jamb filling the river bank to bank – Carried broadside on for half a mile and the boat is nearly crushed in with the pressure. Floe ice is 2 ft. thick and ice freezes on and wedges under the keel so that we can do nothing."

Swannell was not finished surveying in the area. In the spring he had made arrangements to survey a number of lots in the Endako River valley west of Fraser Lake, and he had been given a 50% advance payment.

■ Ferrying horses across the Fraser River at Quesnel. This photograph is taken from the Quesnel side of the Fraser River, and shows the outfit that Swannell's men devised to ferry the horses across the Fraser River.

Swannell spent the first three days of December at Fort Fraser taking care of business. On the 3rd J.H. Gray and his surveying outfit left for Quesnel by sleigh. From December 5 to 7 Swannell and his crew surveyed some lots at the west end of Fraser Lake, and on the 8th they commenced their surveys in the Endako Valley. With the short days the surveyors were only able to work from 9 a.m. to 4 p.m. but despite the snow and winter conditions the lots were surveyed without any difficulty. On December 12 Swannell tied on to Gray's survey and got an observation of Polaris to give his surveys more accuracy. Two days later Swannell completed the Endako Valley surveys, and by the 17th he returned to Fort Fraser.

It was impossible to travel on the river. Fortunately, Mr. Leduke, who ran a small general store along the Endako River, had made arrangements for his friend, Charles Constantineau, to travel by sleigh to Ashcroft, on the CPR railroad, to pick up supplies for his store. Constantineau, from whom Swannell had purchased fresh food in 1908, was glad to have company for his trip, and on December 19 they left Fort Fraser. They traveled on the Yukon Telegraph Trail, the same route that Swannell had taken when he first came to the Nechako Valley in 1908. At one of the bunkhouses where they stopped "a schoolmarm, bound for Fort George, also here for the night, objected to our presence when bedtime came, altho' we had rigged up blankets in front of her bunk to give her privacy. She insisted that we step out into 20º below and remain in the stable while she got changed."

By December 24 the travelers arrived opposite Quesnel. Swannell crossed the Fraser River by canoe and went to complete his financial business with the Northern Crown Bank. The ferry had been pulled out for the winter so the men had to devise their own transportation to get the horses and sleigh across the Fraser River. Two large Indian canoes and a hayrack were procured. "We crossed the Fraser by lashing canoes 6 feet apart and building a railed platform to carry one horse – three trips in all. River full of drift ice." The men stopped at a hotel in Quesnel. Swannell tried to make arrangements for a special Christmas dinner for his crew but the hotel manager wouldn't co-operate so the men traveled with Constantineau down the Cariboo Road.

Spent Xmas night at Shepherd's Road House. Wonderful time here and a marvelous dinner. Old Shepherd on a sofa too drunk to stand; his only commands to his wife being 'Give the boys another drink.' Next morning the womenfolk flatly refuse to accept a cent in payment, but Mrs. S. rather shamefacedly asks [if] I would mind paying the actual cash value of the oats our horses consumed.

After a 400 mile sleigh trip from Endako to Ashcroft the men boarded the train for Vancouver. On December 31, the men arrived in Victoria, the end of a busy and diverse year of surveying for Swannell.

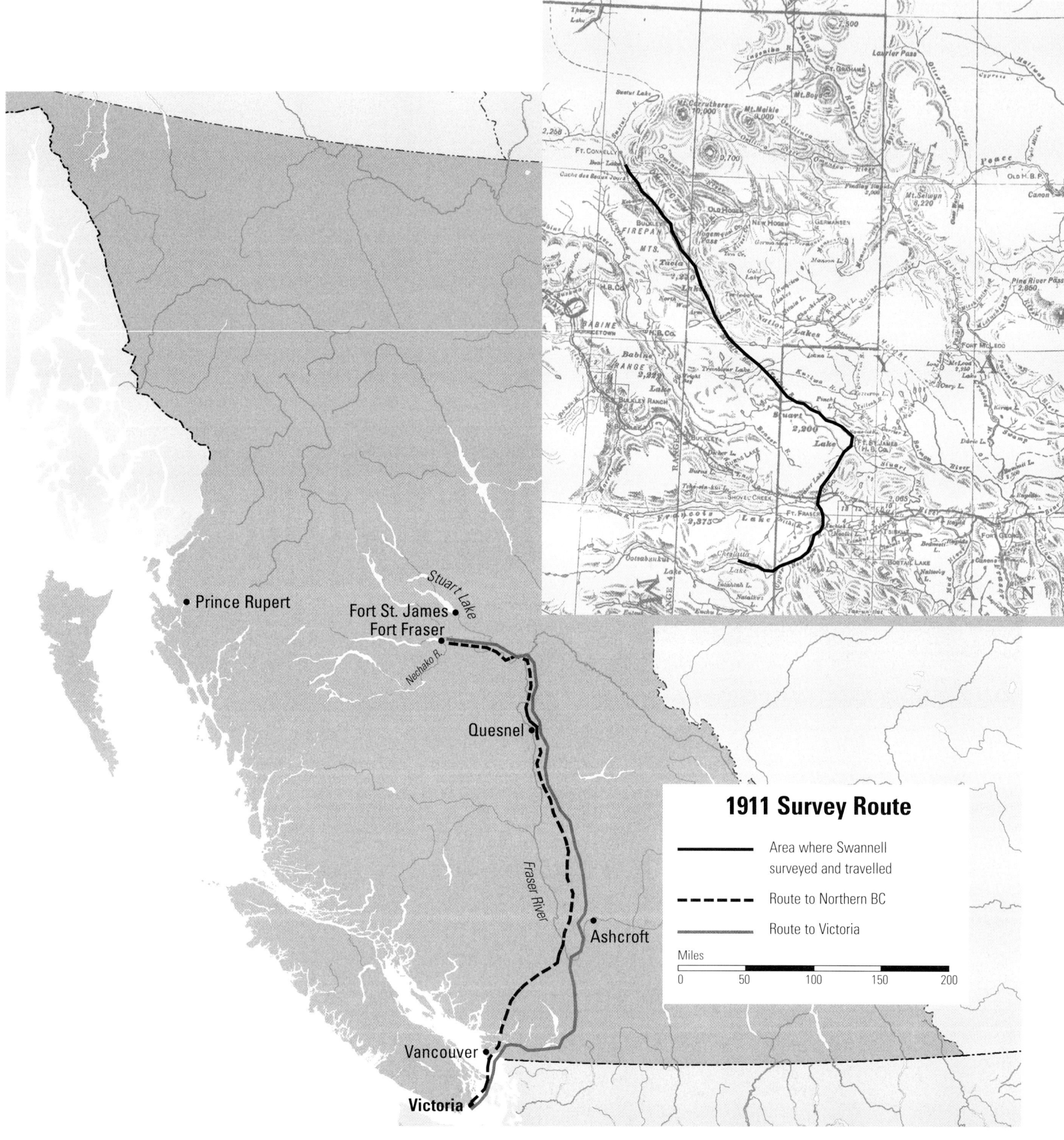

1911

The winter of 1911 was a relatively quiet time for Frank Swannell. There were some small surveys around Victoria. Calculations, maps, and legal work had to be done from the field notes of last year's surveys. In February, Swannell and "Jno. A. Fraser MLA interviewed Lands Minister Ross re Survey Contract. LeDuke incident thrashed out again with Depy. Minister Renwick." In addition to government contracts, bids had to be made for private surveys, and preparations for another field season started. On March 15 Swannell was "notified by HBCo here that a 1600 lb. order has been shipped from Vancouver to Fort St. James." In mid-April Schjelderup and his survey crew left for Lillooet, and Patrick Sharkey, the cook, left for the Nechako Valley to get supplies and equipment ready there.

On April 27 Swannell "interviewed Surveyor-General E.B. McKay re contract on Upper Nechako River. Reserve confirmed and permission to fill in gaps

■ En route to Quesnel on the BC Express Company Special.

SS Fort Fraser - Sestino Rapids, Nechako River. This small sternwheeler is going up one of the rapids on the Nechako River.

outside survey." The Surveyor-General must have been satisfied with the work that had been completed by Swannell and his survey crews in 1910.

In early May Swannell went to Pemberton Meadows for a few days to check on the surveys that Schjelderup had completed in the area. On May 12 "Indian Paul & canoe takes us to Short Portage between Anderson & Seton Lakes. Board steamer." The next day the men left Lillooet, traveling by BX Special Stage. They stayed overnight at 83 Mile House the first evening, and the second night at Murphy's at 141 Mile House. After a three-day journey they arrived at Soda Creek where they boarded the sternwheeler *BX* which took them up to Quesnel.

From Quesnel most of the surveyors traveled up the Yukon Telegraph Trail to the Nechako Valley because there were some lots to be surveyed enroute, one of which belonged to Guy Lawrence, a telegraph operator who later wrote *40 Years on the Telegraph Trail*. Meanwhile, Swannell gave "A.W. Cameron, Northern Crown Bank 16 notes of $500 each for my credit" and paid for 1,000 lbs of freight to be shipped to Fort George. After completing some surveys around Quesnel, Swannell and three men left on the *Fort Fraser* on the 24th of May. By the 27th they arrived at Milne's Landing in the Nechako Valley where Sharkey joined them. That night the *Fort Fraser* tied up at Lampitt's Cache, Stoney Creek. On Sunday, the 28th the men were delayed "at

Stoney Creek repairing boiler tube all the morning. Tie up at Luikarts for night. Run aground at Noonla." The next day, Swannell wrote: "Smash wheel at canyon above Luikarts and hew out new buckets, moving all buckets to inside iron-band. Make to Standing Rock Canyon 3 p.m. – Captain will not tackle it. Unload our stuff. Sharkey & I hit across country for Fraser Lake, arriving 11:30 pm." On May 30 "Sharkey goes to Stella and arranges with Thos Ketlo to bring stuff up in Isidore's boat for $50." They had left some of their goods with Isidore the preceding December. Swannell and William Bunting, the Hudson's Bay Company Factor at Fort Fraser, left for Fort St. James on horseback on June 1. At the Fort Swannell got together two orders of provisions and purchased two canoes. Sharkey and Charles Constantineau brought up packhorses from Fort Fraser. On Sunday the 4th Swannell and Sharkey participated in a friendly rifle match with some of the Fort St. James people. Meanwhile on June 2, Schjelderup and Copley arrived in Greer Valley in the upper Nechako River to complete the surveying they were doing last October.

During June Swannell was busy traveling throughout the Nechako Valley region, organizing and distributing supplies, inspecting the surveys done by Schjelderup and Copley, and working on field notes. On June 19th, while Swannell was at the Greer Valley, "Welch struck on head by a falling tree – and knocked senseless. White in camp with an injured foot. Night frost." The next day

■ Donald Tod family, Fort St. James.

■ Greer Valley survey crew performing gymnastics. Swannell's crew is having fun and providing their own entertainment.

■ "The Scientific Guys." This is part of their Dominion Day celebrations. Swannell took several photographs of his crew in Greer Valley. A few are on the British Columbia Archives website, but most have not yet been published.

Swannell walked 17 miles to the Laketown telegraph cabin. "Thorpe wires for doctor from Nechako but Cotton cannot come down." From Laketown Swannell went to Fort Fraser where he spent a day at the Hudson's Bay Company writing letters and consulting with the Indian Agent, William MacAllan, regarding the Indian Reserve surveys that he was going to undertake once the upper Nechako River surveys were completed.

At the end of June Swannell joined his surveyors for a few days. On July 1, Dominion Day, the surveyors worked a half-day and then celebrated the holiday with "sports and Grand Concert at night." The next day was a Sunday and more sports events were held. It was a day of fun and a break from hard work for the men.

During July Swannell continued to organize the movement of men and supplies. In mid-July Swannell took most of the men up to Fort St. James while Schjelderup remained with a crew on the upper Nechako River.

■ Indian boy dancing, one of the events that occurred during Sports Day. Whenever there was a holiday or a large gathering, people would spontaneously organize some activities. In the remote areas of northern BC this would provide entertainment and a break from hard work and daily routine.

■ The winning team. The boys had a tug-of-war competition followed by the adults.

Sports at Fort St. James, Peter MacDonald, one of Swannell's crew members, is participating in the long jump.

Swannell had a contract to survey Native reserves at Bear Lake. The surveying of these reserves was a co-operative venture of the federal and provincial government. Although the federal government had responsibility for these reserves, British Columbia had jurisdiction for the lands used for these reserves, so they arranged the survey for them. The methods used were similar to those in surveying lots, except on a larger scale. In their instructions the federal government requested that Swannell:

Employ, if possible, at least 2 Indians of each band for whom the reserves are intended. These should be The Chief, one councillor, or men appointed by them. They are to be paid at the rates current in the locality. Separate reports on each allotment or allotments are required. These should set forth the number of men, women, children for whom the plot or plots are intended, their name as a band, the especial reasons for setting the lands apart, their description as to quality, timber, etc.

Bear Lake, one of the headwaters of the Skeena River, had been the site of Fort Connolly, a remote

Hudson's Bay Company post that had operated in the 19th century. To reach Bear Lake Swannell and his men would follow the route that he and Copley had taken in 1909 up the Stuart River drainage. This time he would travel up the Driftwood River and across a portage to the lake. Swannell divided the men into two crews. One was under his charge, while Copley had the responsibility for a second crew that was going to do timber licence surveys at Tremblai Lake. Swannell took three First Nations men: Albert Prince, Jean Marie Prince and Matyaz Prince. William MacAllan, the Indian Agent, accompanied him. Copley took Achille Ketlo, a Native from the Stella reserve, with him.

On July 29, while they were windbound at Fort St. James, Swannell's men participated in a sports day with the people of Fort St. James. Swannell took several photographs of these activities.

Since July 30 was Sunday, the two crews didn't leave Fort St. James until Monday. Despite pulling against a headwind all morning, the men traveled about 30 miles up Stuart Lake. It took them two days to get up the Tache River to Tremblai Lake where they left Copley's crew. "Luckily crossed Tremblay Lake in a calm…Overnight at Surveyor McVitties [LS] camp 4

■ Mrs. A.C. Murray and daughter. A.C. Murray was the Factor at Fort St. James. Mrs. Murray is on the right, while their daughter, Annie, is in the middle.

Indian rifle team at Fort St. James. Albert Prince, Jean Marie Prince and Matyaz accompanied Swannell on the Bear Lake Reserves survey.

miles up Middle River." The next day they reached Takla Lake. For his entry on August 4, Swannell wrote: "Broke camp 6:30 a.m. and reached the forks of Tatla Lake. On account of a headwind and heavy rainstorm camped early. With Pete climbed Mt. Blanchet. Very long walk 5 miles thru timber up a long spur…Left summit 4 p.m., lost our way and reached shore 10 p.m. Drenched." By the 8th of August they stopped for lunch at the remains of Bulkley House at the head of Takla Lake – "only ruins of fireplace left." Swannell and Copley had traveled only a few miles further than Bulkley House in 1909.

Four days of hard poling up the Driftwood River brought the surveyors to Cache des Bonjours, where they left the river to portage to Bear Lake. A.C. Murray, the Factor at Fort St. James, had told Swannell "that the men from Fort St. James bringing up supplies for Fort Connolly met the engages of the latter post here and that there was much interchange of greetings here: hence 'bonjours'." For the men at isolated Fort Connolly the meeting at Cache des Bonjours, which occurred only once or twice a year, was their only contact with the outside.

Swannell and his crew reached the Native settlement at Bear Lake on August 13. He noted "some 6 or 8 cabins at the site of the old post, of which all trace has vanished." His journal entry recorded that, "Father Coccola O.M.I. is here and a large gathering of Sikannis, including Chief Charlie Hunter from Fort Grahame." Father

■ Jean Marie and Maryaz making canoe poles.

■ Tremblai Lake Joe. This photograph shows Tremblai Lake Joe and his family at their village where the Middle River flows into Tremblai Lake.

Coccola, who was the head priest at the Catholic Church in Fort St. James, was on his summer visits to the First Nations people who lived in the remote regions of his diocese. He would spend about a week at each location. Natives from around the area would gather for religious instruction, learning hymns, and catechism. Father Coccola baptized adults and children, and conducted marriages. It was also a time for the First Nations people to visit friends and relatives. For most of the year the Sekannis resided in small, scattered groups north and east of Bear Lake. Some of them lived around Fort Grahame, a Hudson's Bay Company post on the Finlay River. Many of them would gather at Bear Lake in August when the salmon came up the Sustut River to this site.

In his August 13th journal entry Swannell also wrote:

MacAllan & I hold consultation with the chiefs. 'Firs tam Government she come Bear Lake.' Traded 3 plugs tobacco for pair moccasins, 1 plug tobacco and 2 lbs. sugar for beaded mitts. Had a trading agreement with Bear Lake William. Whenever he brought a salmon we chalked a stroke on the cook fly – After five we give him provisions. During a full gathering of the Sikannis one of the chiefs produced a fair-haired little girl and solemnly compared her with Pete MacDonald, also light-haired. Pete had a good alibi having never been within 200 miles of the Sikanni country.

The next day Swannell gave his survey crew a holiday while he "took solar observations, adjusted transit

■ Survey crew poling lashed canoes on Driftwood River. The picture that I took at almost the same location on my trip during the summer of 2004 shows the changes in the glacier and the river during the intervening years.

■ Portage at Cache des Beaux Jours into Bear Lake. This photograph shows Swannell's crew at the historic location where people leave the Driftwood River and begin the portage to Bear Lake.

■ Bear Lake, site of old Fort Connelly. I was able to find this location in the summer of 2004. A comparison with this photograph shows the extent to which the glaciers in this area have receded during the past century.

and consulted with Indians regarding location of surveys." Father Coccola departed for Fort St. James. In his memoirs Father Coccola wrote that he had requested the federal government in 1910 to establish a reserve at Bear Lake. He noted that the area around Bear Lake was rich in minerals, and that prospectors and land seekers were coming into the area. Father Coccola believed that a reserve was necessary to provide some protection for the Native people and ensure that they did not lose all of their traditional land. In writing about his 1911 visit to Bear Lake, Father Coccola noted that the Indian Agent was there along with some surveyors to locate a reserve, although he did not specifically mention Swannell. Mount Coccola, along the western shore of Bear Lake, is named for this well-known priest.

On August 15 Swannell began to "survey Bear Lake I.R. No. 1 which includes Indian village and graveyard." The next day part of his survey ran "along old R.M.P. (NWMP) Trail made in Klondike days." The federal government had originally built the Moodie trail from Edmonton to Dawson City in an effort to get Canadians to travel overland through Canada to the Klondike; however, the few people who used this route found it slower and more difficult than the ship route. In 1906 and 1907 the Northwest Mounted Police built a new section of the trail between Fort St. John and the Yukon Telegraph Trail through Fort Grahame and the Bear Lake area. Both trails would be soon abandoned, but Swannell came across them at times during his surveys between 1911 and 1914.

In his August 17 journal entry Swannell wrote:

■ Departure of Father Coccola from Fort Connelly. Father Coccola is the second person in the canoe and is seated, facing the people on shore.

"Gouvelmen she bling Joy-yaz (little George) – as I help with my small HBCo. axe – so Albert calls it, recalling the story of George Washington & the cherry tree. Take observation for latitude. Albert a rare type – has a strong sense of humour." Two days later Swannell observed: "Salmon beginning to run and Siwashes are making a fish-weir across the river at I.R. No. 3." On Sunday, August 20 Swannell recorded: "All hands in camp except myself. I climb mountain crossed by Fort Grahame Trail and get home 9:30 pm."

Swannell and his crew completed surveying the reserves by August 27 and left Bear Lake on the 28th. They found that the Driftwood River had dropped more than two feet. In several places they struck rocks and had to drop down with poles. On Takla Lake Swannell spent one day surveying a "small Reserve for Daniel Teegee [a Carrier Native] at the Old Landing" and "taking observations for latitude and azimuth (Polaris)." By September 4 they had arrived at Copley's camp on Tremblai Lake, and on the 7th they arrived "at Fort St. James after a hard pull against a head wind. Hearing that Sidney Williams (Govt. Survey inspector) is at Pinchi, we paddle back tonight, camp at 11 p.m. 1/2 mile from Pinchi Village." Sidney Williams (LS) had been hired by the government to travel throughout northern BC to visit the surveyors in the field and inspect their work. Mt. Sidney Williams, between Trembleur and Takla Lake, is named for him. After visiting Sidney Williams the next day Swannell returned to Fort St. James where he spent the following two days fixing up accounts and writing letters.

On the 11th of September Swannell and his men left

■ Indian women at Fort Connelly. Some women have ribbons that must have been from Father Coccola's visit.

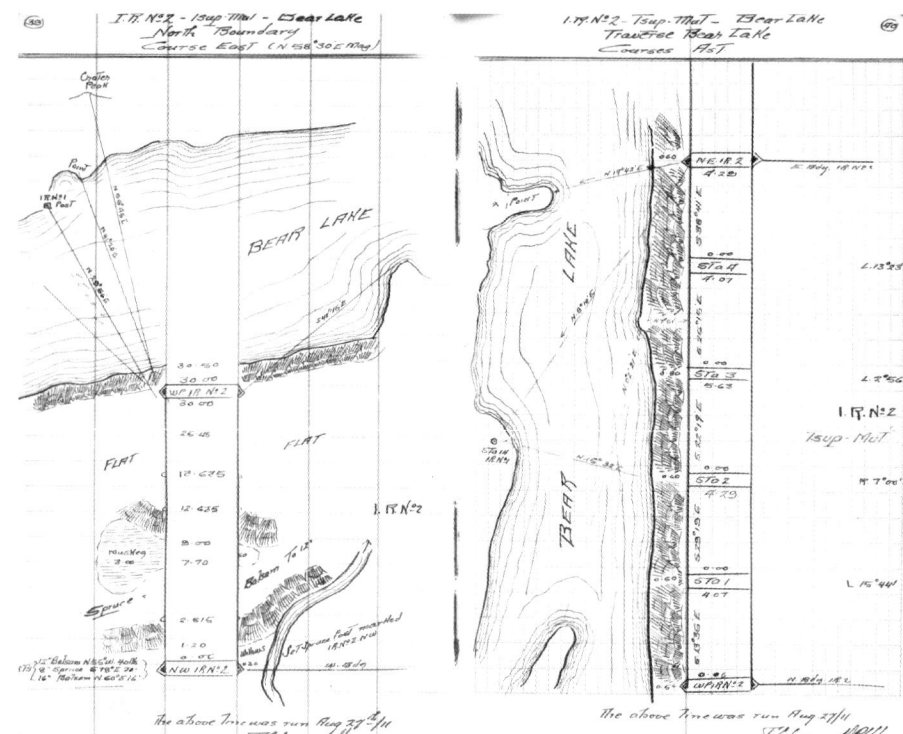

■ This is from Swannell's field book that he used during the survey of the reserves at Bear Lake. The field book is read from the bottom up like the Nechako Valley township surveys. Swannell tied this survey into Bear Lake, Bear River, the surrounding mountains, and the NWMP trail.

■ Sicanni Chiefs at Fort Connelly. Chief Charlie Hunter, from Fort Grahame, is sitting on the ground, the second person from the left. Bear Lake William, who supplied Swannell with fish in exchange for provisions, is standing in the back row, the fifth person from the left.

in a canoe and a boat for Copley's camp on Tremblai Lake. Swannell spent four days working on notes and accounts for Copley's surveys, while his men formed a second crew to get more of the surveying completed. Then Swannell and his crew returned to Fort St. James.

Swannell left Fort St. James on September 21, a federal Election Day. During the next month he did some small lot surveys; returned to Victoria to take care of the finances associated with his surveys in 1911; and inspected some work R.P. Bishop had done in the Lillooet area. By October 25 Swannell was in Quesnel. From there he traveled with two men up the Yukon Telegraph Trail, surveying a few lots. For a few days in early November Swannell went down to Euchiniko and Kluskus Lake for some surveys. There they were caught in an early winter snowstorm. By the 8th of November, when they were back on the Telegraph Trail, the temperature had dropped to –20º F., and it remained below 0 for the next week.

On November 12 Swannell left "Fraser Lake with 3

■ Fair wind and lashed canoes, Takla Lake. This was one of the few days that the men could relax and travel without much work. Swannell's survey crew were on their way back to Fort St. James. William MacAllan, the Indian Agent, who accompanied Swannell is in the front left. MacAllan was Indian Agent at Fort St. James for several years. Before that he operated the telegraph cabin at Bobtail Lake. Swannell took a picture of him there in 1908. Pat Sharkey, the cook, is also in the front, while Pete MacDonald is behind him to the right. Jean Marie Prince is standing in the back, while Albert Prince is seated to the right of him.

■ The Bella Coola Trail. Swannell's crew was traveling along the Cheslatta Trail. Hallett Lake is below them. This is a prominent location on the Cheslatta Trail.

■ On the Cheslatta Trail. Swannell took this photograph of a group of First Nations people from Cheslatta Lake in mid-November. The Natives were traveling from Fort Fraser to their reserve, a distance of about 35 miles. The snow is over the knees of the men. Swannell met these people while he was out surveying along the trail.

horses, packer, Indian Seymour Thomas. Snowing all day. Camp after dark at Chowsunket Lake – no tent and very cold – meat freezes on top while frying in the frying pan." The following day they reached camp near Hallett Lake where Bishop was surveying. During the next week Swannell and his surveyors made traverses on Hallett, Triangle and Bentzi Lakes, all in the same area they had worked the previous fall. On November 18 Swannell took three men with him to survey Reserves at Cheslatta Lake. Before they left they set a triangulation signal on a high hill above Hallett Lake, probably intending to tie the upper Nechako River and Cheslatta surveys together. En route the surveyors traversed "Buntzee Lake (means poor trout) on the ice without using a transit by chaining between shore stations and tying diagonally across the ice. We have our camp on a meadow at a small lake." The next day they were "in camp. Snowed all day without cessation. Horses have good feed in the meadow."

On the 22nd they arrived at Stila-chula on Cheslatta Lake in the late afternoon. There they "hunt up cache in Chief's house & make camp." Early the next morning he got an observation on Polaris and tied into one of his survey stations from last year's triangulation of Cheslatta Lake. He conferred "with Chief Louis & hunt up Indian Agent's post." Swannell hired one of the local Natives, William Charlie, to work on the survey of the Reserves. After the first Reserve was completed, Swannell and the

■ Cheslattas at Stilachula. Swannell took this photograph of the First Nations people at Stilachula. Chief Louis is standing in the middle in the back row. His brother, Baptiste, is kneeling in the front by the snowshoes. To his right, is Skin Tyee, a First Nations person who was well-known in the area. Rose and Sabina are standing on the left.

Chief Louis, Daughters, one of Swannell's few formal portraits. Rose is on the left and Sabina is on the right. They are wearing dress shoes, which is special, but they have a blanket underneath their feet to keep them clean. Although they are only teenagers and the chief's daughters, they have large working, hands.

surveyors "moved camp by horses to I.R. No. 2, 12 miles by trail. One horse played out, substitute bucked pack off twice." On November 30 Swannell tied his surveys into another triangulation station and a survey that E.P. Colley had made in the area. During the first two weeks in December Swannell and his crew surveyed the Reserves at Cheslatta Lake. They also completed a traverse of the Cheslatta River and nearby Murray Lake to add to last year's triangulation survey of the Cheslatta Lake area. The First Nations people lived at the Reserves at Cheslatta Lake until they were displaced by the Kenney Dam project in the early 1950s. In his later years William Bunting, the former HBC factor and store owner at Fort Fraser remembered the Cheslatta First Nations people. "The store did a brisk business during the colourful visits of the Cheslatta Indians who came by horseback and packtrain. They sold their furs, bought supplies, visited and celebrated with their friends. They galloped full bore wherever they went. They were a fine group of people who have now scattered to the four winds."

While they were in the area the surveyors made a traverse of the Cheslatta River to nearby Murray Lake, and they spent a few days doing a triangulation survey of this lake. On the 18th of December the surveyors were back at Fort Fraser. The December 23rd edition of the *Cariboo Observer* reported that Swannell's survey crew had left on the stage south. By the end of December Swannell had returned to Victoria, completing another active year of surveying.

■ Trygve Rognaas made this map from Swannell's surveys of the upper Nechako River. It includes lots that Swannell surveyed as well as his triangulation survey of Hallett Lake and the surrounding area.

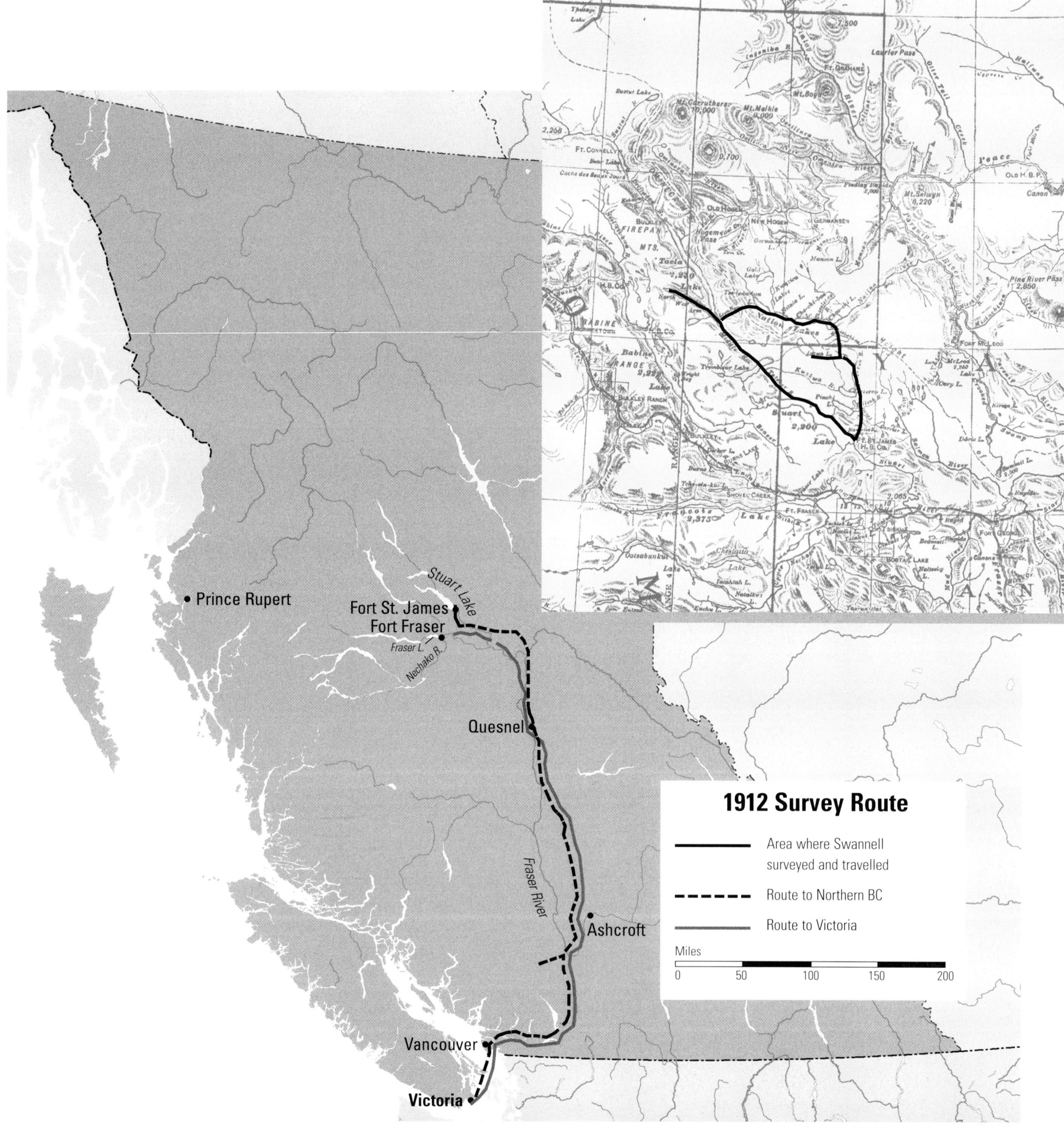

1912

April 19, 1912 – "Surveyor-General to give me season's exploring Nation Lake Region." Little did Swannell realize that such a brief note indicated such a major change in his surveying career. During the past four years Swannell had received government contracts for surveying in northern British Columbia. He had combined these contracts with private surveys to make a full field surveying season. The work had been done in several locations each year. Now Swannell would be spending his field season working for the provincial government on one type of survey in one area of the province.

During the previous four years the British Columbia government had been trying to catch up with the demand for land to be surveyed. By 1912 it had achieved this goal by hiring many additional surveyors on contract, and significantly increasing the funding for the Surveyor-General's Department. Land speculation had

■ Sternwheeler *SS B.X.* in Fort George Canyon. The *B.X.*, the largest sternwheeler on the upper Fraser River, did not have much room to manoeuvre in the canyon.

William Ketlo and family, Nechako Road. William Ketlo hauled supplies and did other short-term work for Swannell on several occasions. Ketlo transported and guided Father Morice on some of his travels.

started to decrease, and much of southern B.C. had been surveyed. The provincial government began to turn more attention to northern B.C. Between 1912 and 1914 the British Columbia government surveyed the 52nd, 53rd, and 55th parallels (the 54th had been previously completed) along with the 124th Meridian to establish surveying control lines in the northern part of the province. At the same time, the provincial government wanted to know more precisely the topography and economic potential of the region north of Fort St. James, and in the Peace River district. To do this they decided to resume triangulation surveying, which had stopped after the departure of Tom Kains, the Surveyor-General, in the late 1890s. The Surveyor-General called this work "exploration surveys," and explained its purpose in his 1913 report. "The advantage of such exploration in advance of regular surveys is obvious, as the information gained in regard to the physical conditions of the district enables the question of further survey to be intelligently dealt with." The Surveyor-General also wrote that, "While many surveys made in the outlying parts of the Province are to an extent, exploration surveys, only two can be strictly classed as such; that in charge of F.C. Swannell in Cassiar and that in charge of G.B. Milligan in Peace River District. The former continued his exploration of last year and extended a rough triangulation over a large tract of country to the north of existing surveys."

There are several reasons that the Surveyor-General selected Swannell to do this exploration, also called exploratory, survey work. Swannell had done triangulation surveys for the government in the Fraser Lake, upper Nechako River, and Cheslatta Lake areas during the past four years. He had spent the previous four field seasons in northern B.C. and knew the area. In particular, Swannell had been to Fort St. James and traveled up to Takla Lake in 1909 and 1911 so he was familiar with the locale where the exploration survey was to begin. Swannell also had a sense of adventure and liked working in remote areas.

The exploratory surveys that Swannell made in 1912, 1913, and 1914 established his reputation as one of British Columbia's prominent surveyors. His success in this type of surveying became the basis for Swannell's surveying career. Gerry Smedley Andrews, who articled under Swannell, and was Surveyor-General from 1951 to 1968, wrote:

> One of the earliest proponents of triangulation after the Kains decade was Frank C. Swannell, B.C.L.S., D.L.S., who has probably done more of this type of survey than any other individual surveyor in B.C. Beginning about 1908 in the Nechako-Omineca region, he conducted exploratory triangulation surveys in many of the heretofore unknown parts of the province, espe-

■ Indians playing lazy stick. The First Nations people are playing one of their traditional games. Daniel Teegee, for whom Swannell surveyed a reserve on Takla Lake in 1911, is in the back row with a pipe in his mouth.

cially the Omineca, Ingenika and Finlay River watersheds, and the vast lake-strewn upper reaches of the Nechako Basin. Mr. Swannell's returns from these surveys comprised more than the usual angle book and austere web of triangles. He made copious sketches of topographic features visible from each station and controlled them by extra "shots" with his transit so that in preparing his official plans he was able to add flesh and blood to his polygonal skeleton of fixed stations in the form of remarkably full topographical detail – lakes, creeks, rivers – cradled in the upholstery of hills and mountains, and indicated by realistic form lines. It is noteworthy, too, that any areas, which were few, that whimsically defied his personal observation from at least one of his ubiquitous stations, were left blank on his map.

Although the British Columbia government lacked precise information about the region north of Fort St. James, the area was not unknown. Small groups of Carrier and Sekanni First Nations people lived in the area, and knew how to sustain themselves in this rugged region. There had been a gold rush into the Manson Creek and Germansen Lake area in the 1870s. There were a few small Hudson's Bay posts, and some non-Native prospectors and trappers. In 1907 Father Morice, the famous Catholic priest who resided at Fort St. James for several years, produced a map entitled Map of the

■ Fort St. James Indians at Dominion Day celebrations.

Northern Interior of British Columbia by A.G. Morice, O.M.I. that was published by the British Columbia government. During his twenty years of travel through the region, Father Morice used a compass, barometer, sounding line, the First Nations' knowledge of the terrain, and his own sense of geography to develop this map. Although Father Morice's map lacked any surveying controls and accuracy, it did provide the first topographic overview of the land between the upper Nechako basin and the upper reaches of the Finlay River. The maps that resulted from Swannell's scientific surveys showed that Father Morice's map was quite accurate.

The BC government wanted Swannell's exploration surveys to have scientific controls and to be accurate, although the standard was not expected to be as high as that for legal surveys. Swannell's main objective was to include as much terrain as possible within the triangulation survey so that the government could produce a map that covered a large area. By exploring as much of the region as possible, Swannell could gather more information about its economic potential and possible transportation routes. Swannell's exploratory triangulation surveys were classified as 3rd Order surveys, which meant that they were supposed to have an accuracy of 1:1000. Over one mile this allowed for an error of 5.28 feet. In the course of a triangulation survey that covered 250 miles there could be a total error of 1/4 mile. Nevertheless, this survey could be used to produce a map that was far more accurate than anything available at the

■ Dominion Day celebration, Ft. St. James 1912. The ladies' sack race was one of the sporting events that occurred during the day.

time. The important element was to include as much detail of the mountains, valleys, and bodies of water as possible. To do so, it was necessary for Swannell to survey from the tops of many of the mountains in the area. Gerry Smedley Andrews wrote:

Swannell specialized in extending accurate survey control over vast areas of unmapped country by triangulation, i.e. networks of huge triangles, the apices of which are located on mountain tops, a method especially effective in the tumultuous terrain of B.C....By triangulation, instead of laborious, costly and inaccurate measuring of distances along the ground, they are precisely computed from carefully measured angles between intervisible stations. A price of this advantage is the arduous and hazardous climb with instruments to the high points and the rigours of inclement weather on exposed alpine sites.

The trips to the mountain peaks appealed to Swannell's adventurous spirit, and the large amount of topographic detail included in Swannell's reports and maps was a hallmark of his surveying work.

George Copley continued to work with Swannell, and from 1912 to 1914 was Swannell's Assistant who was approved by the Surveyor-General's office. Two new men became members of Swannell's surveying crew in 1912, and remained with him through 1914.

Jim Alexander, from Fort St. James, was hired as boatman and guide. Alexander's father had been Chief Factor at the Hudson's Bay Company post in Fort St. James. His mother was a local Carrier woman. His parents must have valued education, for Jim, as a young boy, was sent to St. Joseph's Residential School in Williams Lake. There he met Father Morice. In an article published in The Beaver magazine, Swannell wrote: "The Rev. A.G. Morice, the great linguist, and author of the authoritative History of the Northern Interior of British Columbia, himself told me it was mainly from Jimmy that he acquired his profound knowledge of the Carrier

■ West Takla Lake, taken while Swannell's survey crew was working on the Northwest Arm of Takla Lake.

language." After receiving an education at Williams Lake, Alexander returned to Fort St. James. In the early 1880s Father Morice became the priest in charge of the Catholic mission there. At Fort St. James, Alexander helped Father Morice with his production of his Dictionary of the Carrier Language.

Jim Alexander was the epitome of the rugged outdoorsman. He was big - six feet tall when the average man was about 5'7" - and strong. Alexander had an extraordinary talent for handling canoes and other water craft, and according to Swannell he "could read water like a book." Alexander was also good at handling horses. He was a skilled outdoorsman who was knowledgeable and interested in the human and natural history of northern British Columbia. Alexander's expertise was important for Swannell's success in traveling through northern British Columbia from 1912 to 1914. Swannell and Alexander remained life-long friends. When Jim Alexander died in 1952, Frank Swannell was in Fort St. James. The local First Nations people asked Swannell to be one of the pallbearers at Alexander's funeral.

Nep Yuen (also known as Jim Young) joined Swannell's crew as their cook. As a young man, Nep Yuen had come from China to work on the construction of the Canadian Pacific Railroad. After that he worked as

■ Triangulation Station, Mt. Blanchet. Swannell took this photograph of George Copley at the triangulation station on Mt. Blanchet above Takla Lake. The cairn that they built to mark their station is visible, along with their surveying equipment and rifles that they carried to the top of the mountain.

a tailor in Victoria, logging camp cook, trapper, and placer miner. Nep Yuen had previously worked with Swannell in 1905. Swannell described him as "an excellent cook and general all-round survey hand and horse packer." They maintained their friendship until Nep Yuen returned to China in the 1920s.

Swannell spent the latter part of April preparing for his season in the Nation Lakes area. An order was placed with the Hudson's Bay Company in Fort St. James for supplies and boats. A money advance was once more sent to the Northern Crown Bank in Quesnel. Transportation arrangements were made with the BX for automobile travel from Ashcroft to Soda Creek and sternwheeler travel from Soda Creek to Quesnel. He had repairs to his transit made.

In late April Nep Yuen and Tom Sharkey left with Fred Butterfield (BCLS #87) to make a connection survey near Lillooet that the Surveyor-General requested. Butterfield was affiliated with Swannell in 1912 and 1913, but they did not work on surveys together. On May 21, Swannell, Copley, and Doug Macdougall left Victoria for Vancouver. The following afternoon they departed Vancouver by train, reaching Ashcroft late in the evening. On the 23rd of May Swannell and his men went by automobile to Lillooet where he met Fred Butterfield to review the surveying work that Butterfield was doing. By the 25th Swannell's surveying party arrived at Soda Creek, where they took the BX to Quesnel. The Cariboo Observer reported on Saturday, June 1, that: "F.C. Swannell, the well-known land surveyor of Victoria, accompanied by his party, arrived on Sunday's boat. They are en route to the Fraser and Stewart Lake districts, where they will spend the season. On Monday Mr. Swannell received a despatch summoning him back to Victoria, and left for that city the same day." The telegram Swannell received informed him that his mother was seriously ill. He sent the rest of his surveying crew to the Nechako Valley. Swannell left

■ Frank Swannell at Takla Lake, reading triangulation. Swannell is measuring angles with his transit at a triangulation station along the shore of Takla Lake.

Quesnel on May 27, arriving in Victoria at 7:00 a.m. on May 29. His mother died that morning. The funeral took place on May 31, and the following day Swannell once again departed for northern B.C. At Soda Creek he took the BX to Fort George, and by June 10 Swannell was in the Nechako Valley.

The Surveyor-General had requested that Swannell begin a resurvey of Township 14, R.5 until Butterfield was able to finish his work in the Lillooet area and come to the Nechako Valley. Swannell's surveying crew had already started the work by the time he arrived. The men continued the work, along with a triangulation of Rorison Lake, until the Butterfield party arrived on June 25. The following day, "All hands lay off in place of Dominion Day. Self 1/2 day copying notes."

The next day Swannell and his surveying crew, along with Nep Yuen, left for Fort St. James. They spent Dominion Day at the Fort where they celebrated the holiday with "rifle match & Siwash games." Swannell took pictures of the activities and the people who were there that day.

Before his entry for July 2 Swannell wrote above the date Exploratory Survey. His work on this special government survey was now ready to begin. Unfortunately, a large portion of his field journal for this year is missing. The section from July 10 to September 10 is gone, along with November 3 to December 5. There are a few brief notations, added later, that tell where Swannell was at certain times during these intervals, and many photographs for 1912 have specific dates written on them. In 1912 the Surveyor-General started to include reports from the surveyors on government contracts in the Sessional Papers, and Swannell's report provides some information about his location during the field season.

The first thing Swannell needed to do was establish an accurate surveying base around Fort St. James. In his report to the Surveyor-General Swannell wrote that: "The initial step in making a topographical survey of the

■ Discussing the route to the Nation Lakes. Frank Stephens, the local Forestry official who is standing on the right, and the First Nations people are discussing the best way to reach the Nation Lakes from Takla Lake. One of Swannell's crew is sketching a map in a field book.

Omineca country being to obtain a triangulation base, we were for some days occupied on Stuart Lake erecting a cairn and large signal on Mt. Pope and reading therefrom." On July 2, Swannell recorded that, "Copley, Currie & Macdougall erect signals on small mountain back of the Fort and low point opposite across the lake." On the evening of the 3rd they left Fort St. James and camped near the base of Mt. Pope. The next morning Swannell and his crew broke camp at 3:30 in the morning. "Climb Pope Mt. With 25 lb packs in 1 3/4 hrs. Plane table country in every direction. Taking transit angles referred to Δ 1 & 2 and HBCo flagpole. Descend in 1 hr 5 min after erecting large signal & cairn." Mt. Pope, at over 4,800 ft., is the highest mountain in the Fort St. James area. From this mountain Swannell had a clear view to the north and west for many miles. He was able to see far up Stuart Lake to the area where the Tache River drained into the lake. To the northwest Swannell could clearly see Mount Sidney Williams, a large peak over 6,500 feet, between Tremblai and Takla Lakes.

There were also views of some of the lakes of the area. Swannell used the two signals that had been set near Fort St. James along with Mt. Pope to establish his first triangle. He also used the Hudson's Bay Company flagpole, which was located on a prominent place overlooking Stuart Lake, as a reference point. From the summit of Mt. Pope, Swannell took survey angles to the many topographic features that he could see. Since he would be able to see Mt. Pope from other locations, he took the time to establish a large cairn and place in it a long piece of wood with a large section of fabric attached like a flag. The following day Copley and Macdougall went back to the first two stations that they had established and took surveying angles from them to Mt. Pope. Swannell's first triangle was established. From Mt. Pope the angles that Swannell measured to the features to the north and west would be the first part of the new triangles for his exploratory survey. Although Swannell does not mention it in his notes, he also must have established a baseline distance.

■ Albert's Camp, Upper Nation Lake, is the camp that Swannell reached when he was almost out of food. Albert's family was generous in feeding Swannell's crew and selling provisions to him. A moose skin is visible on the right, and the meat is drying on the rack. The fire in front of the tent is for cooking and warmth.

On the evening of July 5 Swannell and his survey crew left Fort St. James, and began working their way up Stuart Lake. The following afternoon they were at Pinchi Point, which jutted out into Stuart Lake. There they erected a signal, and tied their triangulation into the nearby mountains and islands that they had surveyed from their first two stations and Mt. Pope. The network of triangles was starting to develop. Swannell also noted that he was able to tie his exploratory survey into a survey post on the southwest corner of a nearby lot. In addition, he was able to move his survey forward by taking a reading on Tache Point, another prominent location further up Stuart Lake, that he had sighted from Mt. Pope and Fort St. James. The next day he moved to Tache Point, where he completed his triangulation of this portion of Stuart Lake.

From this location Swannell's crew proceeded up the Tache River to Tremblai Lake. Much of July was spent doing a triangulation of this lake, "as well as establishing main triangulation stations on surrounding peaks." The stations on these peaks would be used when Swannell reached Takla Lake, and some would be visible from the Nation Lakes and Inzana Lake region later in the summer.

At the end of July Swannell and his survey crew reached Takla Lake. Unfortunately, "we found ourselves unable to continue the main triangulation control, owing to smoke from forest fires rendering our previously established mountain signals invisible. We accordingly made a triangulation survey of the North-west Arm of Tatla [Takla] Lake, cruised the country at the head of the arm, and later on erected cairns on, and read from, all

■ Camp at Lower Nation Lake. Johnnie Currie is on the left, Jack Mitchell, the Forest Service representative is in the middle, and George Copley is on the right. This photograph shows conditions in the fall in northern BC.

prominent peaks in the vicinity." It was the best that Swannell could do under the circumstances. At least the triangulation of the smaller northwest arm of the lake, and the nearby peaks, could be added to his exploration survey and would make the map of the area more complete. After his survey of the northwest arm was completed Swannell did some triangulation surveying around the lower end of Takla Lake.

In the last half of August Swannell and his men prepared to move to the Nation Lakes. This was a new area for Swannell to explore and the main reason for his exploration survey for 1912. There were two routes possible into the Nation Lakes. From Takla Landing he could follow the trail that led to the Manson Creek gold mining area and cut a trail for about ten miles into the upper reaches of the Nation River above Tsayta, the first lake. This route would enable Swannell to survey directly down the Nation Lakes. However, this route and the terrain along it was largely unknown. The other choice was to leave Takla Lake near the foot of the lake, proceed up to a pass near Purvis Lake, and then follow the Purvis Lake drainage down to Upper Nation (Tchentlo) Lake, the third lake in the valley. There was reported to be an existing trail. From Tchentlo, however, Swannell would have to survey up to Tsayta Lake and then retrace his steps. Swannell discussed the route into the Nation Lakes with the First Nations people he met in mid-August, along with Frank Stephens, the local Forest Service agent. He decided to go into the Nation Lakes through Purvis Lake. In his government report Swannell wrote that they backpacked a "portable boat and 1,500 lb. of outfit and provisions across country from Tatla [Takla] Lake to Upper Nation Lake. The twenty-mile portage, which was most arduous, took two weeks to make."

The Nation Lakes are on the Arctic watershed. The four lakes are connected by the Nation River. Each section of the river between the lakes is less than five miles

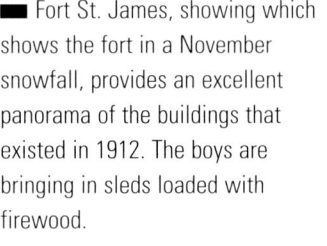
■ Fort St. James, showing which shows the fort in a November snowfall, provides an excellent panorama of the buildings that existed in 1912. The boys are bringing in sleds loaded with firewood.

long. The upper two lakes, Tsayta and Indata, are much smaller than the bottom two lakes, Upper Nation (Tchentlo) and Lower Nation (Chuchi), which are each about 20 miles long. It is about 70 miles from the top of Tsayta Lake to the bottom of Lower Nation Lake. The Nation River flows out of Chuchi Lake and empties into the Parsnip River, one of the main tributaries of the Peace River. At Upper Nation Lake Swannell divided the group. In his report to the Surveyor-General Swannell wrote that he, Copley, and "Mr. Mitchell, of the Forestry Service…triangulated Indata and Tsayta-bat Lakes, and cruised the surrounding country, erecting permanent monuments at practically every instrument station." The second group of men went out to get more supplies and then cut fifteen miles of trail between Lower Nation Lake and Inzana Lake, which Swannell intended to survey later that fall.

Swannell, Mitchell and Copley needed something for traveling on the Nation Lakes. They tried to make a dugout canoe. "Steamed canoe, but in endeavouring to spread it split it from end to end." Instead they constructed a raft. As they worked their way up to the top of Tsayta Lake Swannell recorded the temperature every morning and evening, and took barometer readings to try to establish elevation. At several places he "observed for latitude by merid. [meridian] alt. [altitude] of sun. Observed on Polaris for azimuth." This would be used for establishing the bearing of their survey lines.

By the end of September, Swannell, Copley and Mitchell reached the head of Tsayta Lake where they found a cabin belonging to Babine James. Mitchell and Swannell spent a day "reading signals at head of lake and Copley chopping base on the Peninsula." The next day they "read 6 stations, double-chained base and took Polaris observation at elongation on the base." This gave Swannell an accurate direction, an accurate distance and several survey stations at the top of the Nation Lakes, a key location on his exploration survey.

Swannell had not replenished his supplies since leaving Fort St. James at the beginning of July. When the main part of his survey crew left Upper Nation Lake in mid-September to obtain more supplies at Fort St. James, Swannell, Copley and Mitchell had only 100 lbs. of food. By the end of September, at the head of the Nation Lakes, they had almost run out of food. Swannell noted that beginning September 22 they were eating only two meals a day, fish for breakfast, and grouse mulligan for supper. They tried to live off the land, but the only food that they could find was grouse and fish. On October 6 they reached Upper Nation Lake. "Made down river to Tchentlow (Upper Nation Lake) and found Siwash Albert Michel camped at the lagoon. Purchased beaver meat, whitefish, 4 cups of flour & rice – After Mrs. Albert had built us several bannocks and fed us a muskrat stew. Albert much amused at our near star-

■ Jim Yuen melting snow for water.

vation. 'One time mutehe pretty near mameloose'."

By the 11th of October Swannell, Copley, and Mitchell were almost at the foot of Upper Nation Lake. A "fresh wind and big sea" made travel on the lake impossible so Swannell "went down lake to Milligan's camp at foot of lake." J.M. Milligan, (BCLS #42), did some timber and lot surveys for the provincial government around the Lower Nation Lake and Nation River area in 1912 and 1913. Mt. Milligan, south of the Nation River, is named after him. Swannell used Mt. Milligan for one of his triangulation stations so that he could connect his exploration survey with the lots that J.M. Milligan was surveying in the area.

At the same time Copley left "to hunt up Sharkey, Macdougall & Currie who were left to cut trail across to Inzana Lake." Sunday, October 13, was a "fine calm day. Mitchell & self in Milligan's camp developing and printing photos." The next day the "rest of the crew arrived from inland – having abandoned their trail & made a cache. Decided to go into Inzana Lake by going back on Manson Creek trail." For the next ten days Swannell and his surveying crew ran their triangulation surveys around the Lower Nation Lake and surrounding area. Swannell's field notes started to record snow and cold temperatures. On the 21st of October it was "snowing off and on all morning. S.E. wind. Go down Nation River and photograph country. Hills all hidden in snowclouds." The following day there was "Heavy wind 10:30 a.m. followed by snow-storm in afternoon." On the 27th Swannell crossed "to Nation River by the cut-off trail in order to take photos. 21° frost last night."

Swannell's surveying crew had now reached the foot

■ Lunch at Deep Creek Cabin, Nechako Road. The survey crew has stopped for lunch on the way to Quesnel. Some of the men are wearing coats made from Hudson's Bay blankets again. The telegraph wires can be seen overhead.

of the Lower Nation Lake and were camped along the Manson Creek trail. On the 28th Jim Alexander arrived with five packhorses from Fort St. James. Most of their remaining work would be on land, although they still had their portable boat that they could use on Inzana Lake. As they traveled down the Manson Creek trail, Swannell's crew mapped the trail and took the topography of the surrounding country. Along the way they hiked to the top of Lookout Mountain and set a surveying station there.

In 1873 William Francis Butler had made a winter journey by dogsled through northern British Columbia. He wrote The Wild North Land, a book that related his adventures on this trip. The book was very popular, a classic romantic description of life in the wilderness of northern British Columbia. In the book Butler described hiking to the top of Lookout Mountain and the spectacular view that he saw from this location. He also mentioned a gigantic pine tree that he saw on the south slope of the mountain. Swannell took a photo at this location and made a note referring to Butler and his book. Swannell had a keen interest in the history of northern British Columbia. Whenever possible, he referred to people who had previously visited the same location, comparing his experience with theirs.

On the 30th of October the survey crew reached the "turn off of the trail to Inzana Lake and make 7 1/2 miles up it to Horse Meadow Creek. Tie trail traverse to Pope, Lookout & Murray Mt." Now Swannell's triangulation was complete. From this location and from Lookout Mountain Swannell had been able to make surveying shots that included Mt. Pope, where he had started his

■ Larson's team outside the Occidental Hotel. Swannell's survey crew traveled by sleigh from Fort Fraser to Quesnel. This photograph shows the men in Quesnel at the end of their journey. The Occidental was one of the well-known hotels of the Cariboo.

first triangle in the summer. Over the winter he would be able to calculate his angles and distances to make a large area network of triangles that would extend from Fort St. James up the Stuart Lake waterway to Takla Lake, over to the Nation Lakes valley, down this valley to Mt. Milligan, and along the Manson Creek trail back to Fort St. James through Inzana Lake. In his government report Swannell also noted that at Inzana Lake they were able to tie their triangulation into some of the cairns that they had left on the mountains above Tremblai Lake, giving them another large triangle through mountains besides Mount Pope.

On October 31 Swannell and his men reached Inzana Lake. "Haul canoe over the ice to open water and move 7 miles up the lake. Alexander takes horses to Beaver Meadow 1 1/2 miles down the lake." Swannell completed the triangulation of Inzana Lake in early November, and left a cache of about 125 pounds of food. In a note added later Swannell recorded that: "Benoit Prince agreed to pay for this $25 if in good condition, if not pro rata – this lay in his trapping country."

Swannell's first season of exploration surveying was now finished. In his report to the Surveyor-General Swannell wrote:

Winter having now set in, we proceeded by the Manson Creek Trail to Fort St. James; thence by pack-train to Fraser Lake, from which point we went to Quesnel by sleigh, and proceeding southward, arrived at Victoria the end of November. Our itinerary for the season included the following: Travelled by steamboat, 240 miles; railway, 360 miles; automobile, 440 miles; wagon, 210 miles; sleigh, 160 miles; pack-train 150 miles; boat or canoe, 500 miles; raft, 60 miles; backpacking, 60 miles: a total of about 1,700 miles.

■ Surveyors at 105 Mile House. Swannell,s survey crew traveled from Quesnel to Ashcroft by automobile. Automobiles were just beginning to be used in the winter in the Cariboo. The steering wheel is on the right side.

The Cariboo Observer wrote on Saturday, November 30 that: "Three A-T cars, driven by Rollandot, Boyd and Studebaker arrived here Monday night, and left again for the south on Tuesday and Wednesday. They came up for the purpose of taking out three of Swannell's survey parties." The newspaper also reported on Lawrence Dickinson, who had worked on Swannell's crews during the past three seasons. "A number of the friends of Lawrence Dickinson, who returned here on Wednesday after being engaged all summer with one of Swannell's survey parties, gave a supper in his honor at the Quesnel Café Wednesday previous to his departure for Victoria where he purposes to spend the winter. There were fourteen present, and as the auto left at an early hour Thursday morning they spent the whole night in the jollification, winding up with an oyster breakfast at the Occidental."

In addition to the actual surveying of angles and distances, Swannell was also expected to provide his observations of the land and its economic potential to the provincial government. In his report to the Surveyor-General, Swannell described the existing and potential transportation routes. Swannell wrote that on the Stuart River waterway "the only obstruction to steamboat navigation for this distance is shallow rapids on Tatchi River, only negotiable by steamboat when the water is fairly high." He cited the wreck of the S.S. Enterprise on Tremblai Lake as "proof of the availability of this lake system for steamboat navigation." Swannell recorded that the Manson Creek trail provided good access into the Nation Lakes valley and that "a few deviations from the present trail would give an excellent wagon-road through undulating country the entire way." Swannell described the timber value of the region. He noted that there had been two major fires in the area during the past 40 years, and that "this section has suffered more from forest fires than any other part of British Columbia." He did note a few locations where good stands of timber were left. Swannell described the agricultural potential of the region in great detail and noted the places that he believed could be classified for this purpose. Swannell wrote that, "One remarkable feature, for one familiar with other parts of the Northern Interior, was the absence of summer frost, probably owing to the many large lakes in this region." Swannell observed that:

The only serious attempt at agriculture in this whole region has been at Fort St. James. Here, for over eighty years, the Hudson's Bay Co. have farmed a few acres. They successfully raise all sorts of vegetables as well as oats and barley, and also have a few head of stock which were in fine condition. At the west landing, Tatla [Takla] Lake, an Indian garden was observed, which in spite of neglected cultivation, had an excellent showing of vegetables, including cabbage, beets, carrots, onions, parsnips, and potatoes. A little land has also been cultivated in a desultory fashion at the head of the Northwest Arm, with good results, considering the small amount of work expended thereon. There is no climatic reason why all ordinary field crops may not be successfully raised, the great drawback to the country being the small amount of agricultural land compared with the whole area. Mixed farming and stock-raising would appear to be best adapted to this region.

Swannell noted that, "The whole area explored was almost destitute of big game, with the exception of bear, both black and grizzly, which the Indians said were fairly numerous. We, however, saw not a single one the whole season. There seem to be no deer at all, and caribou were scarce." He did observe that, "The lakes contain very large trout, both rainbow and Dolly Varden, and a large char."

Swannell concluded his report to the Surveyor-General by saying that, "The mass of meteorological data, observations for altitudes, and the detailed topographical notes are being reduced and tabulated as rapidly as possible, and the positions of the trigonometrical stations computed with a view to providing a detailed report and large-scale topographical map. Preparation of this will necessarily take some time."

Swannell's first season of exploration survey for the BC government had been successful. His journal entries for December showed that he and Copley had already started working on the information they had brought back from their 1912 field season on the Stuart and Nation Lakes waterways.

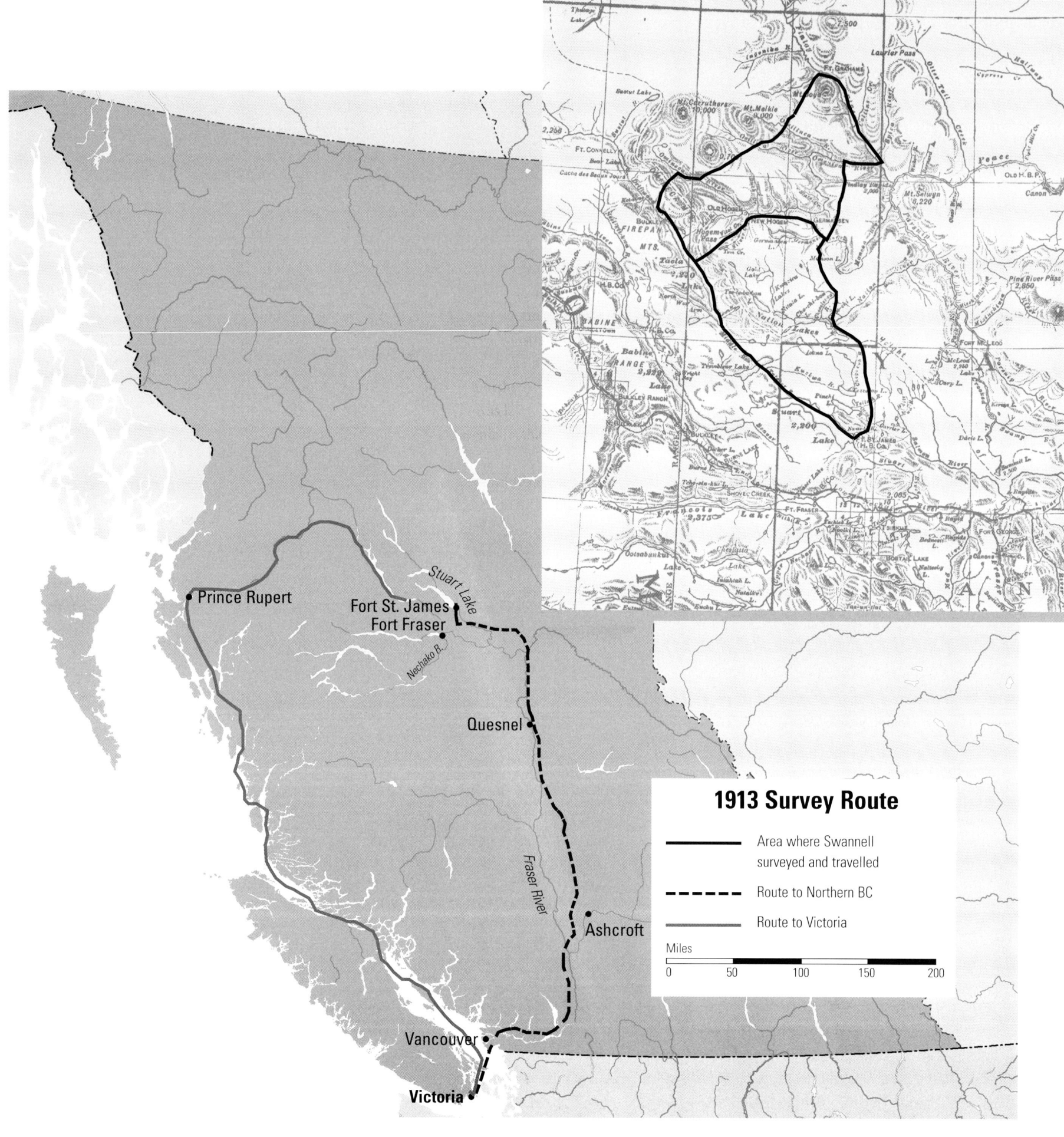

1913 Survey Route

— Area where Swannell surveyed and travelled
--- Route to Northern BC
— Route to Victoria

1913

In late April Swannell received a letter from the Surveyor-General.

I am directed by the Honourable The Minister of Lands to instruct you to continue your Exploration Surveys of last season. I have much pleasure in informing you that your work of last year meets with the approbation of the Department and in view of the success you obtained in these operations it would appear to be hardly necessary to give you definite instructions.

You will continue your rough triangulation to the North and East. There is one matter, however, to which I call your attention. It is neither necessary or possible in a triangulation such as you are making, to obtain results of extreme accuracy, your work being in the nature of an

■ Surveyors on the boiler of the wreck of the *SS Enterprise*. Some of this equipment can still be seen today on Trembleur Lake. This photograph shows Swannell's 1913 surveying crew. Viktor Kastberg, from the Forestry Branch, is on the left. Swannell is sitting on the main part of the boiler, while Axel Gold, also of the Forestry Branch, is behind him. Jim Alexander is standing. Nep Yuen is sitting beside him, while George Copley is standing on the right. The picture was taken by Sam Rosette.

Exploratory Triangulation. Such being the case I must impress upon you the necessity of covering a large area of country rather than looking for refinement in the pursuit of accuracy.

With one exception you are the only surveyor employed by this Department on exploratory work. It is therefore a matter of some importance that the area covered by you be large, the filling in of such detail, such as lake traverses can be done at some future date.

As above stated I leave the arrangement of your seasons operations entirely in your own hands, but I wish you to bear in mind that the Department wish a preliminary knowledge of a large section of country rather than a comparatively correct triangulation of a limited area.

You will be accompanied by a representative of the Forest Branch who will be under your direct instructions in all matters except those pertaining to his work of timber cruising… While you are given authority over this representative of the Forest Branch, I trust that you will have no occasion to excercize same and that relations between yourself and the latter will be such that you will work together…

The Surveyor-General was pleased with Swannell's 1912 exploration survey of the Stuart Lake waterway

■ Kispiox Margaret and baby, taken just after Kispiox Margaret and her family arrived at West Landing. The baby is sitting on the supplies they have unpacked, and one of their horses is in the background.

and the Nation Lakes valley, and the information that resulted from his field work. For the 1913 season Swannell was to extend his triangulation further north and east into the Omineca region. Since Swannell was working in a more remote region and expected to survey more territory than the previous year, the Surveyor-General reminded him several times that he was more interested in the amount of area covered than in a high degree of accuracy. This attests to the quality of Swannell's work in his 1912 triangulation surveys. Forest Branch personnel would again accompany Swannell.

The other surveyor who received a government contract to do exploratory surveying was G.B. Milligan (BCLS #41), who was being sent to the Peace River district. Since Milligan was working far from Victoria, he was going for two years. He would spend the winter of 1913-14 in Fort Nelson, so that he could begin surveying as soon as conditions permitted in the spring. On May 9, Swannell and Copley, who was his Assistant again, left for the Omineca. As in 1910, Swannell's wife, Ada, accompanied him. "Board S.S. Princess May [a CPR steamship] at midnight – Milligan, leaving for Peace River gets great send off from his friends." Swannell would meet Milligan again at Fort St. John in October 1914 when both men had finished their field season.

From Vancouver Swannell, his wife, and Copley took the train to Ashcroft, and then traveled by automo-

■ Kispiox Margaret and baby. The baby is packed in its box cradle and is ready to travel.

■ Plug Hat Tom and canoe crew. This dugout was the Priest canoe, that Swannell hired from Plug Hat Tom while he was doing some of his triangulation surveying on Takla Lake.

■ Bear Lake Tom and family. Bear Lake Tom was a well-known First Nations person in the Bear Lake and Takla Lake area. Swannell estimated that he was born in the 1860s for Bear Lake Tom could remember miners coming through Bear Lake during the Omineca gold rush in 1871. Franklin Pope, of the Collins Overland Telegraph, visited Bear Lake in 1866 and noted a young Tom in the village. Bear Lake Tom is mentioned in Driftwood Valley, Theodora Stanwell-Fletcher's account of her life in the area in the late 1930s. Bear Lake Tom's wife, Sophie is on the left, while their son, Bensen, is between them. The girl is not identified.

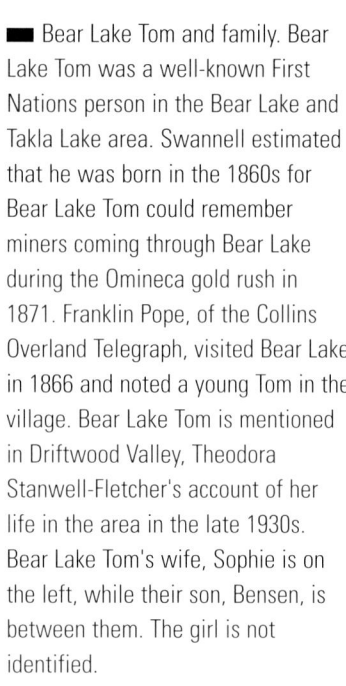

bile to Soda Creek where they boarded the sternwheeler *BX*, arriving in Quesnel on May 13. There, Swannell ordered 1700 pounds of supplies for Butterfield, whose surveying crew would be working again in the Nechako Valley. The following day Swannell re-boarded the *BX*, continuing up the Fraser River to Fort George. He spent the "evening in the big log house of Surveyor Burden" (F.P. Burden, BCLS #9).

Swannell's wife left on the *BX* for Victoria on May 16, and the following day Swannell and Copley began walking to the Nechako Valley. "Copley & self walk on trail 17 miles to 'Slim' Millers and stop over night on floor, worse than on a sand bar." Swannell had visited Slim Miller in 1910 when he was traveling through the area. The next day they walked 32 miles to Cluculz Creek, and the third day they reached Stoney Creek. En route they had lunch at Dad Hills whom Swannell had met during his 1908 township surveys. On the 21st of May Swannell and Copley got a wagon ride to Fort Fraser. Swannell observed the construction of the Grand Trunk Pacific through the community. "Steam shovel and flat cars grading across the town site." On May 23 Swannell and Copley left Fort Fraser at 5:30 in the morning, "across trail to Fort St. James 5:30 pm – a hard hike. Trail very muddy and large patches of snow on the summit. Copley & I lunch at old cabin and hunt down a bushy-tail with 22 pistol – which augments pedometer mileage of trail 4 miles." The next day was declared a "Holiday – we stay with Murray at the HBCo. Visit Sutton's store and buy grizzly and black bear pelt. Shorten drafting table box." On the 27th of May the two "Forestry men A.M.O. Gold & Victor Kastberg and my

■ Babine Trail, McPherson and Daniel Teegee. This picture was taken on the trail between West Landing on Takla Lake and Fort Babine. Swannell is in the middle.

Chinese cook Jim (Nep) Yuen arrive with Larson from Tsinkut Crossing. Camp in the old HBCo schoolhouse." The next two days were spent at the Fort finishing the preparations for the field season. Jim Alexander joined Swannell's surveying crew again, and Sam Rossette was hired "to help up the lakes." Although the last day of May was a Saturday, "made today Sunday. Self at Fort squaring a/c's [accounts] & writing letters. Copley gets a tangent screw made for the transit."

On June 1 Swannell left Fort St. James. As in 1912 his first destination was the summit of Mt. Pope. Although he had an accurate location of the mountain and had already measured angles from there, Swannell

■ Babine, HBCo 1913. The HBC flag is visible in this photograph.

■ Jim Alexander at the site of Bulkley House. This photograph shows the remains of the chimney at Bulkley House.

probably wanted to recheck his starting point. There may have also been specific mountains that he wanted to resurvey. Swannell wrote in his journal that, "Copley, Kastberg & I climb Pope Mt and stay 4 hours on top. Murray says it was named "Pope Cradle" after Maj. Pope of the survey for the Alaska Telegraph Line 1866. Fell asleep on the summit. 1 hr. 45 up, 1 hr 05 down. Got 4 lb. trout." Since Swannell had done a triangulation of Stuart Lake in 1912, the survey crew left the following morning at 6:30 from the foot of Mt. Pope and headed to Tache Point where the Tache River entered Stuart Lake. Swannell resurveyed this station. He noted that he took a reading onto Shass Mt., which is located a few miles south of Stuart Lake. At over 5,800 feet Shass Mt., from the Carrier word for grizzly, is 1,000 feet higher than Mt. Pope. The mountain is visible from many peaks to the north and west.

From Tache Point Swannell's crew proceeded to Tremblai Lake. They observed the "wreck of the 1871 steamer "Enterprise" at west end of lake – boilers, part of heavier framing & machinery." Swannell had already triangulated Tremblai Lake, but he spent a few days surveying from the tops of mountains so that he could add more detail to his map of the area. On June 7 Swannell recorded that, "Copley & I climb mountain S of Mt. William and map a big valley. In windfall came nearly on top of a sleeping black bear, who wakes up and dashes wildly away scared to death." By the 10th of June

■ Ta-wat-enslinstay Rapids, Driftwood River, taken when Swannell's men were transporting supplies up the Driftwood River in preparation for traveling through the Omineca Mountains. Jim Alexander is poling the canoe up the rapids.

Swannell's crew left Tremblai Lake and started traveling up the Middle River. En route Frank Stephens, the local Forest Ranger that they had met at Takla Lake last August, joined them, along with his wife, Betsy Prince.

The following day Swannell arrived at Takla Lake, where misfortune almost occurred. "Snagged canvas boat & narrowly escaped filling and sinking – especially when Gold almost rammed Copley and I when coming to the rescue." On June 12th Swannell "gave Forestry men a months provisions, they to cruise back of the North Arm of Tatla," and the following day Gold, Kastberg, Stephens and his wife left for Cruise Creek.

Now Swannell was ready to begin his new triangulations for 1913. When he was at Takla Lake the previous summer it had been too smoky to do a triangulation of the main part of the lake so he had worked on the North Arm. After the forestry crew left, Jim Alexander and Nep Yuen began setting signals on the east side of the lake, while Copley, Rossette and Swannell worked on the west side. On June 16 Swannell recorded that he had been "37 days from Victoria. Set 6 stations & proceed to West Landing, arriving 6 pm…Babine Siwashes at the Landing." At West Landing the trail from Hazelton and Babine Lake crossed Takla Lake by ferry to East Landing where it continued to the Omineca gold fields.

The next day "Alexander & Rossette leave in red boat for Fort St. James – to meet us at East Landing July 17." At Fort St. James Alexander and Rossette would

■ Pack train horses grazing in a meadow along the Ingenika Trail, taken at their camp near the summit of the Ingenika Trail in the Omineca Mountains.

pick up packhorses and travel almost 200 miles on the Manson Creek trail past Manson Creek to East Landing. These packhorses would be needed for Swannell's surveying in the Omineca region during the summer and fall. Swannell also observed that, "Kispiox Louis with wife Margaret arrives with 10 pack-horses for Manson Creek."

After setting some more survey stations on the west side of Takla Lake, Swannell crossed over to the east side. He noted the construction of a new wagon road to Manson Creek that was part of the revival of mining in the Omineca. If this road had been built the previous year Swannell would have had an easier route into the Nation Lakes. By June 23 Swannell had worked his way up the east side of the lake to Bulkley House at the head of Takla Lake. "An old chimney at the mouth of Bates Creek is on the site of the old HBCo post, and house & stable built 18 years ago by Tom Alexis, now dead – the discoverer of the Tom Creek diggings. Cabins 4 miles down the lake were winter quarters of Klondikers." From Bulkley House Swannell and Copley spent a day surveying up the Driftwood River. In the evening a group of Natives from Bear Lake arrived.

During the last week of June, Swannell, Copley and Nep Yuen surveyed down the west side of Takla Lake, arriving back at West Landing on June 26. On June 29 "Gold & crew stop for lunch." By July 2, Swannell had returned to Takla Narrows, where he had started his tri-

■ Driftwood glaciers.

angulation of Takla Lake almost three weeks ago. He found that "last year Sta. 42 – 3' under water" due to the heavy rain that they had encountered during the past week. On July 3 Swannell observed fresh snow on the mountains. The following day Swannell ran "line 17 chs [chains] to cross a point and finish tie to last season's work." Swannell had connected his triangulation of the main part of Takla Lake with last year's triangulation of the Northwest Arm. The same day Swannell recorded in his journal that "Bear Lake Indians arrive with Plug Hat Tom from Fort St. James."

The next days were rainy and cold, and Swannell spent most of his time working on calculations of his triangulations to check his work before he left Takla Lake. On the 7th of July Swannell wrote that in the evening he took "blue boat to West Landing and bring back 'Priest canoe' and 700 lbs. – all of cache." The following day "Priest canoe hire starts @ $1.00 per day." Swannell hired this big canoe from Plug Hat Tom, who used it to transport Father Coccola from Fort St. James to Bear Lake during his summer ministry.

On July 8th, while Swannell was doing some additional triangulations along Takla Lake, he met "Stephens & wife at Red Bluff," and camped "with foresters one mile above White Bluff. Hottest day yet – south wind. Frost in night." Most of the days Swannell recorded the temperature at 6 a.m., noon, and at 6 p.m. This data was for the barometer that he was using to measure elevation and for meteorological information to present in his government report.

Swannell needed more supplies for the coming months, and this was his last opportunity to make any purchases. On July 12 he left "with McPherson [a former Hudson's Bay Company man who was going to build a store at West Landing] and Gold for Babine along with Daniel Tegee & family." Swannell had surveyed a reserve for Tegee at Takla Lake in 1911, and had met him again at Fort St. James in 1912. Tegee was well known in the Takla Lake area and is mentioned in *Driftwood Valley*, Theodora Stanwell-Fletcher's description of her life in the area in the late 1930s. In a letter to her, Swannell described Daniel Tegee as intelligent and shrewd. Mount Teegee and Teegee Creek along Takla Lake are named for him.

The following day the group arrived "at Fort Babine 11:30 am. Stop at HBCo. Two packtrains arrive from Hazelton." They spent the 14th of July "at Fort Babine making up loads and getting horses to pack supplies back to Tatla. Make arrangements at 2 1/2 cts lb. Rate Hazelton to Babine is 3 1/2 cts…61 days out." By the 16th of July Swannell had returned to West Landing.

The next day Swannell crossed Takla Lake to East Landing. "Alexander & Rossette arrive at East Landing, out 14 days from Fort St. James via Manson Creek with

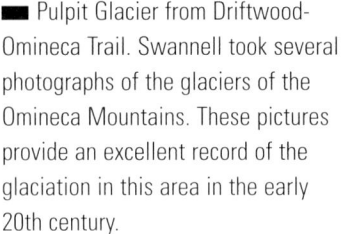

■ Pulpit Glacier from Driftwood-Omineca Trail. Swannell took several photographs of the glaciers of the Omineca Mountains. These pictures provide an excellent record of the glaciation in this area in the early 20th century.

six horses (2 Siwash 4 HBCo.)." The arrangements made a month ago had worked out precisely. Swannell had finished his triangulation of Takla Lake and picked up supplies at Fort Babine, while Alexander and Rossette had brought the horses that would be used for the remainder of the summer.

From Takla Lake Swannell's survey crew made a side trip over the old Fall River Trail to Old Hogem on the Omineca River. In his report to the Surveyor-General Swannell wrote that "the trail was in bad shape, having been seldom used since the seventies. We took pack-train as far as Diver Lake, where we encountered five miles of

■ "Big Kettle," Omineca River. Jim Alexander is about to put their dog, Dick, down into the fumarole. Another crew member is holding one of the dead birds that they found inside.

windfall." In his journal entry for July 24 Swannell remarked: "Gold, Alexander & I climb Diver Mt. (6000') in 2 1/2 hrs from lake. Read topography and take photos. Copley, Kastberg & Sam cutting trail but windfall is so bad have to abandon hope of getting horses over the Old Hogem trail." The following day Swannell "called today Sunday as everybody is played out." On July 26, "Copley & I with 38 lbs each start back packing over the Old Hogem Trail. Bad windfall for 5 miles and almost impossible to follow." The next day the two men lost the "trail in beaver meadows below Quartz Creek; follow Fall River down and reach the Omineca River across a big swamp. Mosquitos awful. Rain most of night…Copley cuts knee badly."

Swannell and Copley finally arrived at Old Hogem on July 28. "Cross Fall River at mouth and get very bad going along an alder-covered hill before we crosscut onto trail. Arrive at Old Hogem 3:30, badly played out. Foundation large store just visible. A new grave, sunk in (of Wrigley's partner). 2 Indian graves and a new dug grave filled in with brush. Mosquitos very bad…Elmer kept store here 1870. Called "Old Hogem" by the miners as he got a corner on provisions."

Swannell and Copley met the rest of their survey party who came in by horse over the Manson Creek trail. Copley returned with the packtrain to Takla Lake while Swannell and Sam Rossette spent about ten days visiting the mining operations in the Omineca. At the Slate

■ Notch Peak Station. Frank Swannell is surveying from the top of this distinctive peak in the Omineca Mountains.

Creek Placer Mine Swannell was "most cordially received by Otterson – who puts me up in his office." He spent all of the next day and evening working on his maps. Swannell also spent a day looking "over old account books and papers of the 1870s in the old Govt. Office. Hudderle & others are pulling the old buildings down and ground sluicing the site." During the last days of their visit to the Omineca, Swannell and Rossette had trouble with their packhorse, Old Bob. On August 6 Swannell recorded that, "Old Bob has strayed off, & only just caught." The next day he observed that, "Old Bob seems sick & very slow." In his entry for August 8 an exasperated Swannell wrote, "Waste 2 hours hunting Bob, who has cached himself."

By August 9 Swannell was back at Takla Lake. During the time that he was in the Omineca, some of the surveying crew had taken part of the provisions up the Driftwood River where they would be working next.

■ Packing to Too-Tizzi Lake. Viktor Kastberg is on the left, Swannell is in the middle, and Axel Gold is on the right. The dog, Dick, is packing the tent. The transit box is visible at the top of Swannell's pack.

They returned on August 10. The following day Swannell called "today Sunday for all hands…Cut drafting table box to fit projection sheets. Take observation at noon for latitude."

On August 13 Swannell and his surveying crew reached the bottom of the Tu-wat-en-indlay Rapids on the Driftwood River where they found the cache of supplies. The next day Swannell and Jim Alexander went "down below cache and get Bear Lake Charlie's canoe. Put 1200 [lbs] in canoe, when 700 more than enough for size of canoe & stage of water. Drag up rapids, one place having to make a channel among the boulders… Sappoose (sheer) in one chute – nearly break canoe & spill load. Gold & Copley back, climb escarpment, mapping – Kastberg ahead clearing trail."

■ Moving camp down the Stranger River. This is the raft that Swannell's crew built to take the equipment and some of the men down the Stranger (Mesilinka) River. Partway down the river the raft had to be abandoned because of log jams and the swiftness of the river. Jim Alexander is on the raft at the left.

■ Swannell Party at Fort Grahame. Swannell and his survey crew, their dog, Dick, and two Hudson's Bay Company employees are in this picture.

About eight miles above the Tu-wat-en-indlay Rapids Swannell's survey crew found the Omineca/Ingenika trail. Swannell described the trail in his government report. "Leaving the Driftwood at Mile 25 from the site of Bulkley House, we crossed over to the Omineca by the Ingenika Trail. This trail was originally a Sekanni foot-trail over which horses were taken during the Ingenika excitement. It crosses numerous muskegs and early in the season is almost impassable for packhorses. At the summit are very large meadows affording excellent feed. It was observed during the season that large high-altitude meadows characterize all Omineca passes."

After three days of travel through boulders, muskeg and windfall Swannell's crew reached the summit of the trail. The next day, August 19, Swannell, Alexander and Copley climbed "Nep Yuen Mt. and triangulate. Set up cairn 6 ft. high. 'Sunday' for rest of crew…All beyond Omineca River high mountains." Swannell was now in the middle of the Omineca Mountains, and saw many high peaks around him. To the north and west lay many mountains and valleys that he had not seen before. On August 20 Swannell and Copley climbed "Peggy Peak (6065') and make tie back to Tatla Lake. Take solar observation. Top very narrow, slide-rock and over-hanging cliffs, small glacier – country to Bear Lake grass plateaux…Fine warm day. Name peak after our little dog – badly scared and whimpering on the narrow summit." Swannell now had his triangulation of Takla Lake tied with his survey of the Omineca Mountains, and his solar

■ Warehouse at Fort Grahame. The men enjoyed an opportunity to read newspapers and learn about what was happening in the outside world.

Survey Crew Making a Canoe Near Fort Grahame. Jim Alexander, in the front, supervised the construction of this dugout canoe. Ross, the HBC Factor at Fort Grahame, is observing the work. There are a lot of wood chips around the canoe.

observation enabled him to calculate his latitude. Swannell spent the next day calculating his observations and measuring a short baseline. He now had an accurate location, distance and triangle at a key point of his 1913 exploration survey.

By August 22 Swannell reached the Omineca River. Although the 23rd was a Saturday, "weather uncertain so make today Sunday." The following day, "Alexander, Rossette & I go seven miles up trail to the Big Kettle." Swannell's government report contained a detailed description of this unusual geological feature.

The "Big Kettle," at Mile 308 3/4 (from Fort St. John) on the Police Trail, was examined and photographed. It is situated close to the Omineca River on a large creek. The "Kettle" itself is at the top of a conical mound about 25 feet in diameter at the base and 15 feet high, and is a vent or fumarole of 6 feet in diameter and perfectly cylindrical, filled to within 5 feet of the rim with a soft red earth resembling hematite. Many small birds, mice, several bushy-tailed rats, and a large owl were found dead in the bottom. We had been previously been told by Sikanni Indians, who stand much in awe of this place, that birds flying overhead are mysteriously killed in mid-air. Shortly after our arrival the fumarole emitted strong puffs of sulphurous gas, which, however lay heaviest in the bottom of the vent. It quickly stupefied our dog and compelled us to clamber hastily out of the fumarole to avoid suffocation. I am unable to account for the presence of so many dead birds in the bottom of the fumarole, other than by accepting the Indians' statement as correct; as it seems probably that, at times, the emission of gas may be very much stronger and less intermittent than during the period of observation by me. About an acre around the "Kettle" is built up of a spring-deposited rock resembling travertine. Many mineralized springs seep out, forming stagnant pools and oozy patches of reddish and yellow mud.

After visiting the Big Kettle, Swannell's survey party proceeded down the Omineca River until they reached the Northwest Mounted Police Trail between Fort Grahame and Bear Lake. Swannell had been on this trail in 1911 near Bear Lake. In his government report Swannell describes the trail. "The Police Trail from the Omineca to Too-tizzi Lake runs north-east, following a path through a broken range of extremely rugged glacier-bearing mountains, whose summits are over 7,000 feet in altitude. The trail so called is execrable, the pass (4,000 feet) being choked with huge boulders alternating with patches of muskeg – all the detritus of the slowly receding glaciers." Along the route to Too-tizzi Lake Swannell and Copley climbed "Ferris Mt., both peaks. Get tie back to Olson and Peggy Peaks. Large glacier across Ferris Creek." In the meantime Gold and Kastberg went across the valley where they climbed "Axelgold Peak and set a signal flag."

At Echo Lake, before the pass, they established a camp where Swannell set up and measured a short base. "Alexander gets a 90 lb. goat, badly wounds another which gets away as he ran out of cartridges." The next day Swannell and Copley climbed "Notch Peak (7100') & Echo Mt. and fix triangulation points ahead." Meanwhile, Gold and Kastberg were continuing their forestry work "cruising up alpine valley to the northward." On the last day of August Swannell finished "readings onto triangulation points from the base." The camp at Echo Lake was another important location for Swannell's exploration surveys. He had set up another short baseline, and visited the tops of several high peaks. From these he was able to tie his triangulations back into the Omineca Mountains and forward into both the Fort Grahame and Manson Creek areas where he would be surveying during the coming months. Since the peaks were so high and rugged, Swannell was able to take sightings onto several mountains from each location, so he would have many calculations to make accurate triangulations. At this camp Swannell also found the Mile 280 marker for the Police Trail.

On the first day of September Swannell's survey party headed over the pass through the mountains. They encountered a "snowstorm, weather raw and cold…It took horses 4 hours to go 3 miles and Bob nearly kills

■ Indian Graveyard, Fort Grahame.

himself." By September 4 the survey party reached Too-tizzi Lake. "Move camp to Too-tizzi Lake (Long water) in one trip, all hands packing their dunnage & dog Dick the big tent. 1600 lbs all told (1100 provisions & kitchen outfit). Make a large raft to avoid packing over trail along lake." The next day was a "rainy day – made it Sunday. Put decking on raft…Snowing hard in mountains." Swannell spent the next few days making a triangulation of the lake. "Gold & I climb Too-tizzi Mt. – Very cold wind and driving snow. Clouds prevent sight back. Arrive in camp 6:30 very tired."

From Too-tizzi Lake Swannell's survey party proceeded down a small river draining out of the lake until they reached the Mesilinka [Stranger] River. They decided to make a raft and run the Mesilinka to save time. On September 12 Swannell wrote: "Copley, Alexander & I run 17 miles down river. Very swift & bad drift-piles. Run into one and under a sweeper and 'sappoose' (sheer) 3 times. When swept under sweeper we jump over as it passes but dog Dick is swept overboard – makes for shore and will have no more rafting, thereby showing sound common sense." The next morning Swannell climbed Mica Mt. to make some triangulations. Despite their misadventures the previous day, Swannell decided to continue rafting down the Mesilinka River. "Run down raft loaded with 1600 lbs down to where the trail finally leaves the river – several bad driftpiles and I have a narrow escape. A log-jamb ahead almost blocks the river. I jump ashore to snub the raft, am swept off my feet and would have been ground between raft and gravel bar if Jim had not grabbed me." In his government report Swannell wrote that: "We sent our horses down the trail "light," and ran our outfit twenty-five miles on a raft, but this mode of transport proved to be very hazardous."

From the place where they left the rafts Swannell and Copley walked on the Police Trail "through to Fort Grahame 22 miles, arriving 3 p.m. Find two poling boats had just left for Fort McLeod. Receive hearty welcome from Postmaster Ross." The following day their packtrain arrived. While they were at Fort Grahame the men, led by Jim Alexander, "cut down cottonwood and commence making a dugout canoe." The Surveyor-General must have told Swannell that he would be continuing his exploration survey into the Finlay River area in 1914.

In the book, *Finlay's River*, R.M. Patterson describes the construction of this dugout, based on interviews with Swannell and Copley. A section of the cottonwood tree over 20 feet long was sawed off, and Alexander marked the design of the canoe on the tree. They needed an adze but there was no adze, forge, or blacksmith tools at Fort Grahame. Alexander and Copley, aided by Nep Yuen and Rossette, built a hot fire of cottonwood bark. They used the fire to heat metal and make the tools that they would need. Then took a Hudson's Bay axe and shaped it into something similar to an adze. With this improvised adze and their axes the men hollowed out the cottonwood log. Then they filled the part that had been

■ Lining up Black Canyon on the Omineca River, a prominent location on the Omineca River. Jim Alexander is pulling on the line, while one of the other men is poling the canoe.

hollowed out with water, and placed in it big river stones that had been heated in the fire. As the water got hotter the wood became soft and pliable. Under Alexander's direction the men spread the log and shaped it into a canoe. Then they cut several thwarts and placed them crosswise into the canoe to hold its shape. Gradually the temperature was lowered and the wood shrunk, tightening against the thwarts. Some final shaping was made, and their canoe for the following season was ready. While Swannell had been unsuccessful in building a dugout on the Nation Lakes in 1912, Alexander's expertise produced a canoe that was heavy but would serve Swannell's surveying crew well on their 1914 survey of the Finlay River.

After their canoe was completed, Alexander and Swannell decided to try it out on the Finlay River. "We two run down the Finlay for 5 hours and camp 33 miles down at head of the deadwater." The next day they continued down to Finlay Forks where the Finlay and Parsnip Rivers met to form the Peace River. Swannell met several people at this small settlement. "Find Sicannies Antoine & Pierre camped 200 yards from us last night. Two prospectors trade bear & moose meat for tea. At the Forks Louis Petersen (Dane, came in at the McConnell Creek Rush.) Harrison & Wilton, the Babine Fire Warden & partner – the last two 13 days from Manson Creek on a raft." On September 23, Swannell recorded that: "We take Sunday – for fun run the Finlay Rapids and half-fill in the big waves. Run hard into right bank of the Peace, but careful steering is need-

■ Summit Moody Trail, Wolverine Range, Omineca, where Nep Yuen became temporarily lost in a snowstorm.

ed. Peace River is 1/4 mile wide at the Forks, the Finlay 300 yds. Mrs. Adams & Petersen to afternoon tea. Arrange with Mrs. Fox (Indian) to fix up marmot robe for me. Warm sunny day."

The next day Swannell and Alexander headed back up the Finlay River to the mouth of the Omineca. They proceeded by canoe up the river to the Black Canyon, a well-known landmark on the river. In his journal Swannell again refers to *The Wild North Land* and William Butler's experience at that location. From the Black Canyon Swannell and Alexander walked to the Mesilinka River. In two days they found Copley and Rossette chopping a trail along the river and bringing the packhorses from Fort Grahame. By October 1 Swannell's survey crew had reached the junction of the Mesilinka and Omineca Rivers, and on the 4th they were at Black Canyon where Swannell and Alexander had left the canoe. The following day Swannell and Alexander took the canoe through the canyon.

From Black Canyon Swannell and his men crossed the Omineca River and headed to Manson Creek on the old Moody Trail. They spent the first night at Muscovite Lake. "Lost kitchen horse with Sam – nothing but hardtack for supper. P. Rasmussen, trapper, at the lake with a cabin." The next day the men had a "long hard move to Wolverine Summit (4500'). Very steep final pitch. Very cold on summit and horses 'all in'. Camp in clump of stunted spruce." The next day it was "blowing a blizzard in the morning. Trail across summit only a stick marker every quarter mile or so. Nep Yuen gets lost on top in the snow but gets out by backtracking. Very bad trail, muskeg at foot of mountain and on the flats the old trail hardly traceable." In a note added later, Copley wrote: "Chief went back to locate cook, but missed him & come on to us in none too good a humor." On October 9 the survey crew reached the Omineca mining area at Slate Creek. "Breakfast 4:30 – Cold night 1° above zero…Meet Child (Otterson's caretaker) and camp in a

■ Manson Creek, Omineca, showing some of the buildings and tailings at this mining community. George Copley is on the left. Ed Sullivan, Max Gebhardt, and Billy Steele, well-known Omineca miners, are next to him.

cabin. Otterson had told me he would leave a small keg of port wine for us to help ourselves from. An hour after our arrival Child came into the cabin & ostentatiously carried it out. But Nep had forestalled him by filling 2 camp kettles – the miners told us later Child said port wine was a tipple for gentlemen not surveyors."

The following day Swannell and Copley went down to Manson Creek, explored the old mining buildings and visited some of the miners. That night, "we stage a banquet for all of them – including Child." On October 11 Swannell spent the morning working on his map. "In afternoon go into Manson with Steele and overhaul papers in the old Govt. office. Ah Hoo arrives. Take latitude observation." The next day Swannell went down to the mining community of "Germansen with Ah Hoo, placer miner here in 1871."

On October 13 Swannell and his survey crew left on the Manson Creek Trail for Fort St. James. They camped that night near the summit of the trail. The following morning they started early, "but make poor time, 6" snow on trail, but very little near Gillis Grave. Old head-piece "Memory of Hugh Gillis a native of Souris, Prince Edward Island. Died 19th Aug. 1872 aged about 30 years. According to Alexander Gillis had been handed a letter here – sat on a rock Jimmy showed me, read the letter – tore it to shreds and shortly after blew his brains out. The packers pieced together the scraps of paper and found it was from his girl who had written him announcing her marriage to another in very callous terms."

By October 17 Swannell and his crew reached the Nation River where they had been last fall after their triangulation of the Nation Lakes. J.M. Milligan was continuing to survey in the area and Swannell went to his camp. The following day there was a snowstorm so Swannell spent the day with Milligan. On October 19 Swannell continued down the Manson Creek Trail, heading for Fort St. James. They passed Cataline's camp and climbed Lookout Mountain again. "We found I think Butler's gigantic pine tree – only it is a fir! A snowstorm drove down on us & we could read into nothing but Pope Mt." Fortunately, Swannell was able to tie his triangulation into his starting location, as he had been able to do in 1912.

On October 22 Swannell and his men arrived at Fort St. James, after almost five months in the field. "Camp in Donald Tod's old house. Surveyor McElhanney arrives." [W.G. McElhanney, BCLS #38 was surveying the 124th Meridian near Fort St. James.] The next day Swannell called "today Sunday for all hands. Developed photos and started on accounts. Visited Father Coccola & Sutton in the afternoon."

During the next few days Swannell did some small surveys around Fort St. James including a pre-emption for Jim Alexander, one for Sam Rossette, and another for a clerk of the Hudson's Bay Company. On the 28th of October Milligan and his survey party arrived at Fort St. James, and on the 30th the two surveyors walked to Fraser Lake, staying "overnight in the telegraph cabin with Olaf and Louis Larson. Meet Butterfield at Trout Creek." The next day Swannell went to the Fort Fraser

■ Ah Hoo, a miner of 1871, a Chinese miner and one of the original miners to come to the Omineca gold fields at the beginning of the rush in 1871. When Swannell met him in 1913 he had been in the Omineca for 42 years. Ah Hoo was born around 1850 so he probably came to Canada as a teenager and worked at Barkerville before moving north during the 1871 Omineca gold rush.

townsite and arranged transportation to Burns Lake. On November 1 the packtrain arrived from Fort St. James. The following day the "survey team defeats town at football. Concert in evening."

On November 3 Swannell, Copley and Nep Yuen left for Burns Lake on a four horse team. En route they stopped to visit Leduke, whom Swannell had met in 1910 during his Endako Valley surveys. During their trip Swannell observed the construction of the Grand Trunk Pacific. On the 4th they stopped for the "night at Shovel Creek – very poor accommodation. Road very bad, contractor's teamster can only haul 1000 [lbs.] with 4 horses." The next day they got "through to Freeport at end of Burns Lake and by badly overloaded launch to Decker Lake. Bad accommodation at Traveller's Hotel. Steam shovel and 'dinky' engines working day & night on the G.T.P. Grade. End of steel only one mile out." On the 7th of November Swannell and his survey crew left "Rose Lake by mixed train 9 am, arriving at Smithers 5 pm – average speed about 10 miles per hour – delayed by a car off the track. Smithers full up, so walk on 3 miles to Chicken." The next morning the men left at 7:45, reaching Prince Rupert at 5:30 p.m. That evening they boarded the *SS Prince Rupert*, a steamship operated by the Grand Trunk Pacific,, arriving in Vancouver on November 10 and reaching Victoria at 7 am the following day. That same day Swannell called on the Surveyor-General. The following week Swannell noted that he had been "instructed by Surveyor-General to prepare a map of the whole region covered in 1912-13 for insertion in Report."

In his report to the Surveyor-General Swannell was

GTP construction at Decker Lake.

optimistic about the agricultural possibilities of the land north of Fort St. James. "The Finlay and Parsnip Valleys contain by far the largest area of undeveloped agricultural land in the Northern Interior." Swannell estimated that there was 600,000 acres of good agricultural land in these valleys. "An entire absence of the dreaded summer frost until late in September is one very favourable sign. This fact was specially noticed by us in the past two seasons, and all prospectors we met verified our experience…The fertility of the soil is vouched for by a luxuriant growth, especially noticeable in the sub-irrigated cottonwood river-bottoms." Swannell did note that there had been very little farming in the region. "No farming has been done as yet beyond a little vegetable-gardening at Fort Grahame and McLeod, and this year at Finlay Junction. At Grahame good potatoes as well as other vegetables and raspberries and gooseberries have been grown over a considerable period of years. Swannell concluded his observations by stating: "I have no hesitation in predicting a great future for the great Finlay-Parsnip Valley. Agricultural development will result in an impetus to the mining industry."

Swannell commented on mining in the Omineca district. "The old placer-workings at Manson, Germansen, Vital and Tom Creeks were visited. At present placer-mining is being done on Manson and Germansen Creeks and some quartz-mining near the Fall River…Much ground known to be auriferous will remain unworked until transportation facilities are improved. At present it is very difficult to get supplies or machinery in from outside, every pound having to come in by pack-horse or toboggan."

Swannell observed the difficulty of accessibility into the region. He noted that most of the trails were in poor condition and that there was no regular river transportation. This made the cost of goods very expensive and hindered the development of agriculture. He cited the settlement at Finlay Junction as an example of the potential and difficulties of the region. "At Finlay Junction two stores were started this year and over a dozen pre-emptors have acquired holdings. A most important settlement will, I am certain, soon centre here owing to its strategic position at the junction of the three great rivers, the Peace, Parsnip, and Finlay. At present the settlers are handicapped by having to bring in their supplies by way of Giscome Portage from Fort George. The placing of a steamboat on the Upper Peace would, I am convinced, result in this section settling up very rapidly. A trail, or better, a wagon-road to Manson Creek would provide an immediate market for the first settlers."

1913 was a successful year for surveying for Frank Swannell. He was out in the field longer than 1912 and had visited new territory. He traveled on many of the trails and hiked to the top of several mountains. Swannell took many photographs of the people and places he visited. He had completed the first exploration survey of this part of northern British Columbia. During the winter Swannell would be busy calculating his field notes and making his triangulations, and the first scientific map of northern BC would emerge from his surveying and sketches.

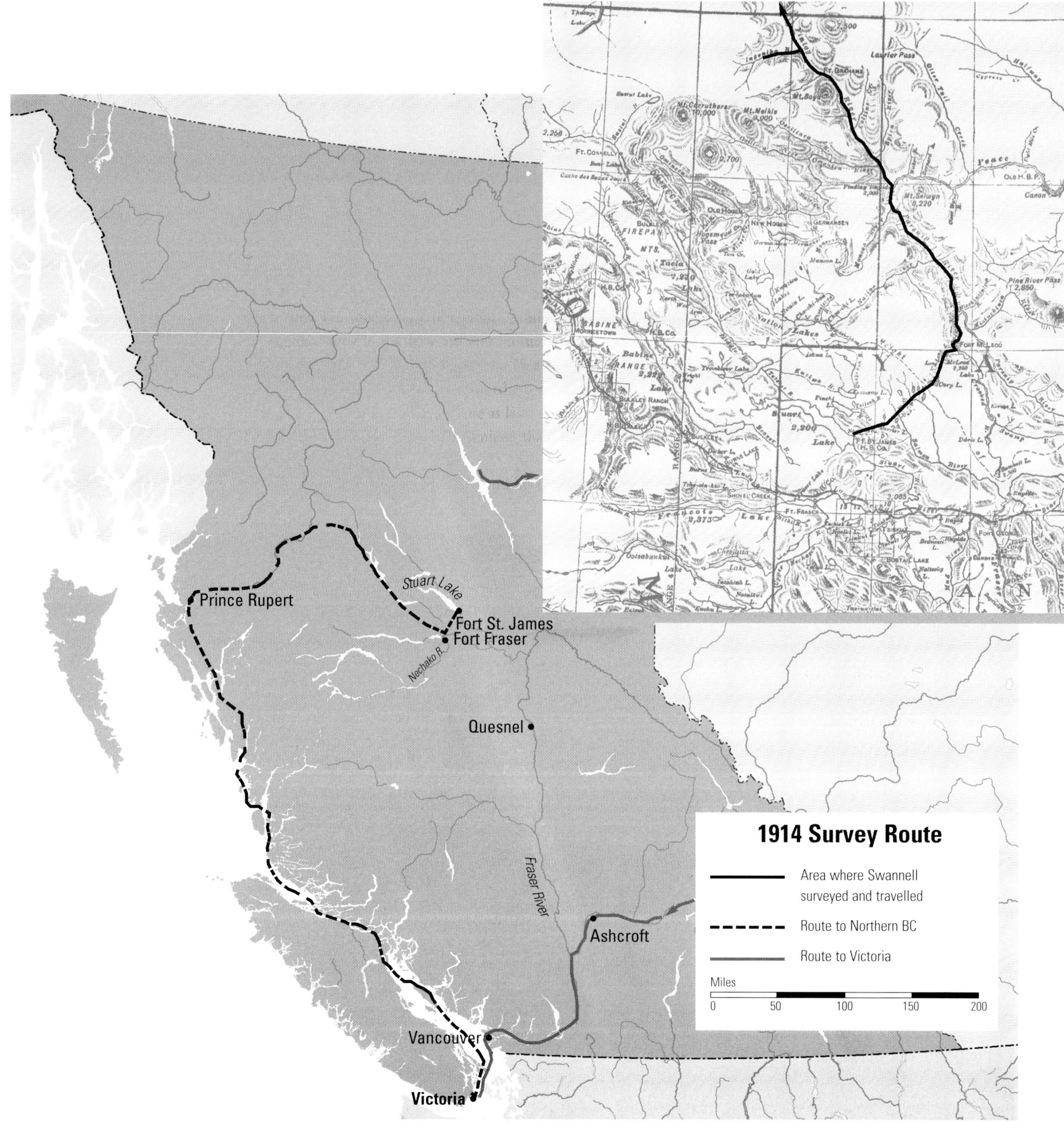

1914

On April 22, 1914 the Surveyor-General wrote Frank Swannell:

I am directed by the Honourable Minister of Lands to instruct you to continue your explorations of last year in the Northern part of the province…

In view of the lack of detailed knowledge as to the conditions of the country to be examined by you, it is inadvisable to attempt to give you detailed instructions.

I may say, however, that your work during the past two years has been extremely satisfactory to the Department and, under the circumstances, I feel satisfied such instructions are unnecessary, though I may, again, ask you to bear in mind that the accuracy of your triangulation is of less importance to this Department than that you should cover a large amount of territory.

I would suggest that, as you are commencing work in the vicinity of the Findlay and Parsnip Rivers, you leave some pre-arranged mark which can be tied on to by Mr. McElhanney, who will complete the survey of the 124th Meridian in this immediate vicinity…

It would be extremely satisfactory to the Department could you extend your exploration

■ Leaving Fort St. James for McLeod The schoolhouse where the men stayed is the building on the right, while the fish cache is on the left. Both buildings can still be viewed at Fort St. James National Historic Park.

Indian Village, Fort McLeod.

Fort McLeod, meridian altitude for latitude. The Native people are inspecting Swannell's transit while he had it set up to take some readings on the sun to establish latitude at Fort McLeod.

so as to make a tie with the base-line carried on during the past few years by surveyors in the Ground Hog section, but I do not, for a moment, wish you to seriously interfere with your Summer's programme in order to make this connection. There is at least another year's work in the north country and should you think it preferable to leave this connection till the year 1915, you are at liberty to do so.

You will employ Mr. G.B. Copley as assistant at a salary of $125.00 per month.

I have the honor to be, Sir, Your obedient servant

G.H. Dawson, Surveyor-General

As the Surveyor-General noted in his 1913 report: "The region north of the Railway Belt and extending

from the summit of the Rocky Mountains on the east to that of the Cascade Range on the west has for the past five years practically monopolized the attention of the public, and possibly over 80 per cent of all land surveys, both Government and private, has been in this section." Most of the provincial government's surveying had concentrated on central and southern BC during the previous few years. However, Swannell and other surveyors had started surveying northern BC, and through their work the geographic features of this part of the province were becoming accurately located and the economic potential better known.

1914 was an important year for surveying in northern BC. G.B. Milligan continued his exploratory survey in the Fort Nelson area. T.A. McElhanney extended his survey of the 124th meridian from the 55th parallel to the foot of the Rocky Mountains. T.H. Taylor continued his survey of the Groundhog coal basin in the Cassiar, while Frank Swannell expanded his exploratory survey further north into the Finlay River.

A new dimension was added to Swannell's surveying for 1914 – longitude. It is relatively easy for surveyors to locate the latitude of any of their stations through observation of the sun's position in the sky at its highest daily angle. Longitude is based on the location of a position relative to 0º longitude, the Greenwich Meridian. To calculate longitude a person needs to measure the sun at its highest point in the sky, the zenith, and know the exact time. A chronometer, a specialized timekeeper, is used for accurate recording of the time. Since the sun travels around the earth, 360 degrees longitude, every 24 hours, it travels 15 degrees every hour, or one degree every four minutes. The chronometer must be set based on the exact time of a location whose longitude has been precisely located. By measuring the sun at its zenith and recording the time on the chronometer it is

■ Working up the Finlay at high water. In high water, when the water was too deep to use poles the men pulled on branches or pushed on logs to work their way upstream.

possible to calculate the longitude of a location.

In 1912 W.G. McElhanney, with the assistance of an astronomer from the Dominion Observatory, located the 124th meridian at the 52nd parallel. The May 25, 1912 edition of the *Cariboo Observer* reported on the work of Mr. McDiarmid, from the Dominion Observatory, and noted that the time signal he was using came from Field, a town along the CPR and telegraph line near the BC-Alberta border. During that year McElhanney and McDiarmid also established monuments at the 53rd and 54th parallel. [The Sinkut Astro pier, which marked the intersection of the 124th meridian and 54th parallel is displayed at the community museum in Vanderhoof.] The following year T.A. McElhanney surveyed the 124th meridian up to the 55th parallel where he established a survey monument. This meridian passed about 11 miles east of Fort St. James. Swannell took a chronometer for the 1914 field season. Since he was traveling through Fort St. James again, it would be easy for him to go to the 124th meridian to take an observation for longitude and set his chronometer.

In the summer of 1913 the Chief Astronomer of Canada sent two observing parties to BC. One of the six locations where they found the precise latitude and longitude was at the 6th Cabin on the Yukon Telegraph trail. In his 1913 report the Surveyor-General noted that the position at "the 6th Cabin has been connected with the base-lines run in the Groundhog basin…The use to which the telegraph line has been put – namely the establishment of a geodetic position in connection with the survey of extensive coal areas in the wilds of northern British Columbia – was probably not contemplated by the original promoters." The longitude of the 6th Cabin was given as hours, minutes and seconds (to 1/100 of a second) in relation to the Greenwich Meridian.

In his instructions to Swannell the Surveyor-General asked him to leave a well-marked survey station that McElhanney could survey to from the 124th meridian. This would enable all of Swannell's work from 1912, 1913 and 1914 to be connected to McElhanney's line and give a precise longitude and latitude for the geographical features that Swannell had surveyed. The Surveyor-General also hoped that Swannell could connect his survey with the Groundhog coal basin surveys that now had a precise location since they were connected to the exact longitude and latitude of the 6th Cabin.

■ Tracking up the Finlay River. When the water was shallow and there was a clear bank for walking, the men used long tracking ropes to pull the canoes up the river.

Swannell's 1914 exploration survey had the potential to provide a bridge across northern BC between the 124th meridian and the 6th Cabin on the Yukon Telegraph Trail. This would enable a large portion of northern BC to be accurately located. However, it was unknown whether Swannell would be able to reach the headwaters of the Finlay River in one field season or whether T.H. Taylor would be able to reach the divide into the Finlay watershed in 1914. The Surveyor-General did tell Swannell that his survey link could be completed the following year. As in 1913 the Surveyor-General mentioned that covering a large amount of territory was more important than accuracy.

Swannell's 1914 exploratory survey of the Finlay River and its tributaries is famous in the history of northern British Columbia. It is featured prominently in R.M. Patterson's book, *Finlay's River*. Patterson used Swannell's journals and had several interviews with Swannell and Copley. Swannell wrote an article, "Ninety Years Later", for *The Beaver* magazine in 1956. In this article Swannell described his experiences on the Finlay River, and compared it with the epic 1824 trip of Samuel Black of the Hudson's Bay Company. Several books carry brief descriptions of Swannell's 1914 trip.

Although the scope of Swannell's survey was simple, it was a difficult venture. Almost no one lived on the Finlay River above Fort Grahame, so the survey crew would have to be self-reliant. The Native Sekanni traveled through the upper Finlay region but did not have permanent residences there. R. G. McConnell of the Canadian Geologic Survey had traveled up much of the Finlay River in his 1893 exploration. Some of the people trying to reach the Klondike passed through this area, and a few miners prospected in the region. However, only one expedition, led by Samuel Black of the Hudson's Bay Company in 1824, had traveled by canoe from the mouth of the Finlay to its source at Thutade Lake and back.

Swannell took a small but veteran crew in 1914. George Copley returned for the sixth season and was Swannell's assistant once more. He was also directed by the Provincial Botanist to make a collection of the plants and seeds that he found in the Finlay watershed. Jim Alexander was their boatman and guide again, while Nep Yuen returned for another season as cook and survey helper.

Throughout the beginning of 1914 Swannell and Copley worked on their calculations and the maps based

■ Taking out the Finlay River fur, HBCo boat. On their way up the Finlay River Swannell met Factor Ross of the Hudson's Bay Company taking out the furs he had bought during the winter.

on their 1913 field notes. In mid-April Swannell sent in his final voucher for the 1913 field season. On the 25th of that month Swannell "filed with Chief Geographer Aitken all inch to mile sheets, referenced map photo card & topo & triangulation books. Get balance of stationery packed." One surveying year was finished, and another was ready to begin.

Two days later Swannell, Copley, Nep Yuen, and Swannell's wife and son, Lorne, left on the *SS Prince Rupert* for Vancouver. Swannell and his family spent the evening visiting friends in New Westminster, before he and his crew boarded the steamer again at midnight. Swannell had decided to travel to northern BC through Prince Rupert, using the same route as last fall since that had been faster and more convenient than traveling through the Cariboo.

On the 29th of April Swannell and his crew arrived in Prince Rupert in a rainstorm. He immediately called on "General Super Mehan, [the Grand Trunk Pacific's Superintendent] & present Surveyor-General's letter. Mehan denies sending down a wire that the road would be open to Priestly on Monday 27th. Leave Rupert 10 am, arriving at Smithers 7 pm. Considerable building but an execrable townsite, all spruce swamp." They left Smithers the next morning, "arriving at Rose Lake 7 pm. The railway road-bed is very dangerous on account of high water in the Bulkley River and subsidence of the grave. The cars get derailed twice. Find our baggage left behind at Smithers. By bribing a car-sweeper get berths in Colonist Coach overnight. Wire to Davidson, Rupert to give section boss authority to run us down in a hand-car." On the first of May, "leaving Copley to round up the baggage, Jim (Nep Yuen) & I go on horseback to McKennes Road House at Burns Lake, 18 miles. Trail

■ Sikanni chief Charlie Hunter, Fort Grahame. Swannell had met Charlie Hunter at Bear Lake in 1911. Charlie is photographed with his family.

very muddy, the road impassable." The following day "Dick Carroll & I start down Burns Lake in a boat but cannot get past the big rock-cut on account of ice. Abandon boat & walk to Freeport & start for Priestly. No go. Meet two men poling a push car along the flooded grade and return with them to Burns Lake – arriving 11 pm…Below Freeport finding the wagon-road flooded and the corduroying gone we had taken to the grade, but our two leading horses carry about 20 feet of the embankment into the river and are nearly drowned." On May 3, while they were still at Burns Lake "Mehan wires refusing to let us have a handcar through. Copley wires that he cannot get down."

The next day, while Swannell was still stuck in Burns Lake, he received "news of train wreck & wire from George [Copley] that he is unhurt. Chief Engineer Van Arsdol & several others badly injured. Last car overturned into Bulkley River." In a note added later Copley wrote that he, Schjelderup and the train agent were on the back of a car and managed to escape unhurt. After the wreck Copley and one of the engineers built a big fire with new railroad ties. Copley, a talented storyteller, and the engineer "kept the spirits of the passengers up by swapping yarns all night. Engineer tells a tale about a box-car that wrecks a whole train. Next morning a box-car train arrives to take the stranded passengers out, but on account of the engineer's tale the passengers refuse to board. The train pulls out & later the engine comes back with a string of flat cars on which the passengers crowd."

By the 5th of May Swannell was able to leave Burns Lake and walk through to Priestly. "Arrange for packhorses for Copley and boat to go up Decker Lake. Road bed from Freeport very bad – the Endako River lapping the top of the grade & the embankment too soft to bear even the weight of a man. One engine completely ditched at Priestly." The next day Swannell continued walking down the railroad "grade to Stella, not stopping at Endako, where extensive yards are being laid out, but no permanent buildings of any description." On the 7th of May Swannell finally reached Fort Fraser. "At the bluff

■ Mooseskin boat, Fort Grahame. Charlie Hunter is beside the frame of a boat that will have a mooseskin hide put over it. George Copley is on the left and Nep Yuen is on the right.

2 miles up the lake [Fraser] a fill being made in 2 1/2 feet of water suddenly goes through the roof and disappears into 35' of water. Fort Fraser absolutely dead except for the presence of railway & bridge crews…All along the line the breakdown of railway communications results in a scarcity of food." Swannell spent two days in Fort Fraser, resting, and visiting friends, including William MacAllan and George Ogston. The second morning was spent "at new railroad bridge with MacAllan. 3 concrete piers built in caissons, about 225' span. River [Nechako] rising rapidly and liable to stop construction of middle caisson. Railroad now carried across on a piling-bridge on a 'shoo-fly spur'." On the second afternoon he went to the Hudson's Bay post where he was "hospitably received by the Buntings."

On May 10 Swannell walked to Fort St. James where he stayed with Murray, the HBC factor. Two days later Copley arrived, and on the 13th Nep Yuen came. "We camp in the schoolhouse." It had taken sixteen days for Swannell's crew to travel from Victoria to Fort St. James, and several days of valuable surveying time had been lost.

Swannell spent almost a week at Fort St. James. He bought supplies for the summer and sent them by pack train to Fort McLeod, took care of his accounts, and made two small surveys. Saturday, May 17, Swannell walked "out to camp near McElhanneys line (124th Meridian) 11 miles from Fort St. James. On Sunday he spent "most of day taking observations for time on McElhanneys line. Cannot get any observation of sun's meridian transit as sun is obscured." Swannell must have sent someone into town to send a telegram to the Surveyor-General requesting the latitude for the station that he was using for his survey. The next day a "wire from S.G. brought out by an Indian lad" gave him the information that he needed.

On the 19th Jim Alexander's time started and Swannell's survey crew left Fort St. James headed for Fort

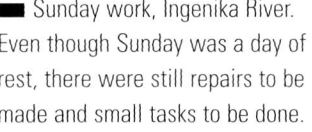
■ Sunday work, Ingenika River. Even though Sunday was a day of rest, there were still repairs to be made and small tasks to be done.

McLeod. Swannell began taking some of the measurements that were part of the routine of their survey. The temperature was recorded at 5 a.m., noon, and 5 p.m. Swannell also started comparing the chronometer against the time on Swannell's, Copley's, and Nep Yuen's watches in case any of them would be needed later for recording time. Their two barometers were observed at 5 a.m. and 5 p.m. in an effort to establish accurate elevations.

By the 22nd they arrived "at McLeods Lake 10:30 am. Attempt to cross horses at River Mouth but nearly drown our white horse in a flooded islet…At the regular trail ford the water is very swift & deep and landing too steep. Camp inside HB enclosure." The next day Nep Yuen and two Natives left in a Hudson's Bay Company canoe, taking about 1300 pounds of food to cache at the mouth of the Ingenika River above Fort Grahame. Swannell spent the day taking "observation at noon for latitude. Obtain 54º 59' 12". In afternoon observe for time. Spent afternoon figuring out observations and making up accounts."

On May 26, thirty days after departing from

■ Crew at Windlass Portage in Deserters Canyon. Jim Alexander and Nep Yuen are hauling the heavy dugout canoe up a hill in Deserters Canyon, using a Spanish windlass they have built.

G.V. Copley at lobstick signal on Paul Mountain. The lower branches have been cut off this tree so that it can be used to identify this mountain when surveying from other peaks. Lobsticks were sometimes used as signals, particularly when there was only one large tree in a location. The transit box is next to Copley.

Victoria, Swannell, Alexander, and Copley left Fort McLeod in a heavily laden canoe, and began paddling down the Parsnip River. By supper the next day they were at the small community at Finlay Junction where the Parsnip and Finlay meet to form the Peace River. The following day Swannell took more observations and sent off his accounts while the rest of the "men made canoe poles and nailed a strip along the gunwale for strength and to avoid swamping."

Swannell's crew began paddling up the Finlay River to Fort Grahame on May 29. High water on the Finlay River made for difficult paddling. Swannell described their difficulties in "Ninety Years Later": "There was no 'poling bottom' and the current at bends ran like a millrace. The 16-foot poles gave no purchase, with only two or three feet above water slatting against the gunwale. Lining was impossible. We dragged ourselves upstream by the overhanging willows; maybe the bowman could drive the spike of his pole into a jutting log and pull the canoe up inch by inch or the sternsman could in turn shove against the log. Some of the channels through log jams were so swift that, as Jimmy [Alexander] phrased it, it was 'jes lak going upstairs'." En route they met "Fort Grahame Factor Ross going down with books and furs in a big clumsy boat. Meet Indian Aleck Pierre & Klootch and get a little salt in exchange for tea, Copley having forgotten our salt. Make to 2 miles below Collins House. Jim A. says Collins was a Fort Grahame manager long ago who, in the late fall, going up in a scow with all his supplies got stopped by ice and had to build a house and winter – all trace of the house has vanished." One of their lunch stops was at Vital's Bar, where Vital LaForce found gold on the Finlay River.

By the 2nd of June Swannell's crew arrived at Fort Grahame where they found "about 60 Sikannis camped awaiting the priest. River below Grahame is very badly cut up with bars and sloughs but we get considerable lining. A new expression "Hudson Bay honey" – syrup cum bacon grease." During this day Swannell had caught up with Nep Yuen and the two Natives in the canoe. The next day Copley and the two Natives chopped out "a base line on the flat below the graveyard." Swannell invited the Sikanni Chief, Charlie Hunter, whom he had first met at Bear Lake in 1911, for lunch. The next two days Swannell and his crew spent "overhauling outfit, calculating observations & measuring base line." This base line was tied into Swannell's 1913 surveys that ended at Fort Grahame.

On June 6 Swannell's crew began their exploratory survey of the Finlay River. The next day they measured another "base to fix peaks ahead and sketch river. River in many channels full of bars – shifting every year over a mile of country." By June 8 Swannell reached the Ingenika, one of the major tributaries of the Finlay, flowing in from the west.

The main purpose of Swannell's exploratory survey was to map as much of the Finlay River drainage as possible, so Swannell needed to travel up the major tributaries like the Ingenika. Swannell's crew set up a base camp about 1/4 mile up the Ingenika River and cached most of their supplies there. Then they started their survey of the Ingenika River, using their dugout canoe. The Ingenika was high because of spring runoff, and it was slow navigating up the swift river. In two days they traveled twenty miles reaching a large branch of the Ingenika that would later be named the Swannell River. At camp that night Nep Yuen promised the men rabbit pie, but they pretended not to understand him. "Jim, the cook, amuses us by remarking 'you no savee labbit! All samee pussycat no more tail.'" On Sunday, June 14 the men took a "holiday. Finish base readings & take observations for azimuth and latitude. George & Jimmy go hunting." A few miles further up the Ingenika, Swannell noticed placer mining that was about twenty years old and a cabin that was almost undermined by the river. In his June 17 journal entry Swannell observed that the "river is rising rapidly and full of float timber. Has risen a foot a day for nearly a week." By the 22nd of June Swannell estimated that they were about 50 miles up the Ingenika River.

Swannell's journal entries recounted the hardships of ascending the Ingenika River. At the same time the men had to survey the river and valley. There was a trail near the river that Swannell used for a survey traverse. Sometimes they would climb a nearby hill or mountain to triangulate the mountains of the area. Occasionally a base line would be set up and observations for latitude would be made. R.M. Patterson, in *Finlay's River*, described evening life in camp for Swannell's crew while they were on the Ingenika:

> And in the evenings, until daylight fails, Swannell will be busy with his calculations and his field notebooks – Copley helping him or working on his botanical specimens. They work after the wind has dropped and right through the hour of the mosquito. They carry on until they can no longer see the small neat figures or the detailed sketch plan of some intricate bend of the river with its islands, its confining mountains and its tributary streams. The fire smoke wreathes and eddies around camp, keeping the mosquitoes more or less at bay; the homely clatter of pots and pans indicates that Nep Yuen is on the job, making ready for the morrow's breakfast. Jim Alexander is not idle: he is sharpening the axes, splicing a line, replacing a broken pole with a slim sixteen-foot fire-killed stick from the bush.

On June 24 Swannell's crew left their camp fifty miles above the Finlay River. By mid-morning they reached a cabin and cache at fifty-three miles. Swannell decided to make a base camp there. He noted, "Above

Cascade Canyon, illustrating the difficulties of getting the heavy dugout canoe through the canyons on the Finlay River.

here the trail is only a trapping blaze." From this camp Swannell, Copley, and Alexander took a three-day trip up Wrede Creek, which flowed into the Ingenika from the southwest. After this the entire crew followed a trail across the summit on the north side of the Ingenika and reached Tutachi Lake. There they saw a moose which Copley and Alexander shot. The next day, while Alexander and Nep Yuen dried the moosemeat, Swannell and Copley climbed Espee Mountain and made a triangulation survey of the area.

In 1914 Dick, the dog, once again accompanied Swannell and his survey crew. Early in the field season Swannell's crew picked up a stray dog that they named Caesar. Caesar had a bad habit of stealing food. Jim Alexander had previously tried to break Caesar of this habit. According to Copley, "Jimmy had shot a coyote pup on a Sunday – skinned a hindquarter, fixed it up nice and threw it to Caesar – who promptly started to bolt it down as usual, but vomited it up when he found it to be dog meat." However, this had not stopped Caesar from stealing food, so at their camp at Tutachi Lake the men decided to once more try to break Caesar of this habit. In a note added later to Swannell's journal Copley wrote: "At this camp after cutting all the meat we wanted from the moose we tied the tramp dog Caesar to the carcass and left him there till we came back from the mountains, so that all he had to do was eat and drink. When we got back he had had such a feed of meat that he would never steal anymore." Patterson also described this incident and wrote that: "When they returned to camp in the evening they found a bloated caricature of a dog, distended and abject, stretched flat out on the grass as far as it could get from the moose carcase."

After returning from their second trip, Swannell's crew moved up the river on July 4. "Very hard poling all day. At Mile 56 there are several channels all but one choked by log-jams. The river now a succession of riffles. Nearly get through one rapid with the pole, but are swept back and almost swamped. On the North Bank high rugged mountains timbered down to the river.

■ ISurvey crew drying moosemeat at Big Bend on Finlay River. This was after Jim Alexander killed the moose in midstream. The men have constructed a drying rack, made a frame for the moose hide, and built a saw horse. There's smoke from the fire.

Camp 35 is 65 miles up the river." After a Sunday in camp Swannell's crew moved one mile up the river to an "old placer camp at foot of a ditch. Bad log jam & rapid, in which we nearly swamp the canoe. At old burnt cache next the ditch blaze – June 20/1899 Please do not molest this cache Jos Malzer – three more illegible names." On July 7, "Copley and I climb Bad Luck Mt. (6200) leaving camp 5 am. Weather turns bad, rain below and snow and sleet above all day. Get no sights at all. Shelter under overhanging slab of rock and build fire with heather but are frozen out and have to go down to timber line on the north side – 6 hours rain here but stay on top until 4 pm. Bad route, hair-raising going down & get stuck in an ever worsening chimney & have to climb back." The next day was spent "in camp all day, no observations to be had." The following day Swannell and Copley climbed "Bad Luck Mt. again, 2 1/2 hrs to summit 3700 above camp, the last 2000 feet up an interminable rockslide. See 8 goat in the crags and shoot 2 but one is inaccessible. Snow squalls at 10 am, but weather not bad afterwards. Make triangulation ties. The valley ahead is nearly everywhere burnt. Another bad climb down – 1200 feet of loose slide and then get cramped between crags in a couloir. Pack 100 lbs goat meat into camp."

Swannell's crew had almost reached the end of navigable water on the Ingenika. On July 10 they "moved upstream about 6 miles to a big eddy. River very bad, a succession of rapids and drift-piles. See the wreck of a large clinker-built boat on top of a log-jam. Put together entirely with wooden pins – no nails. One wonders as to the fate of the crew and how long ago it was." There was one more move to a camp that Swannell estimated to be 78 miles up the Ingenika River. There he found another old cabin and cache. From this camp he climbed two peaks to take triangulations. The mountains that surrounded him would later be named the Swannell Range.

At this location, far up the Ingenika River, it was only about 40 miles by land to Thutade Lake, the headwaters of the Finlay River, and the ultimate goal of

■ Foot of Fishing Lakes on the Upper Finlay River.

Swannell's exploratory survey. However, Swannell was supposed to map the Finlay River and reach the lake by this route. It was time to return down the Ingenika.

On July 14 Swannell's crew began their descent of the river. "The worst place was above Camp 35 which we were foolish to attempt to run…Our average speed 8 miles per hour for the first 20 miles, current averaging 6 mph. We were delayed nearly an hour at the Big log jam, having to line, there being no possibility of dropping down by pole, deep water having scoured out under the logs." In two days they were back to the Finlay River. Swannell recorded that, "Ascent of Ingenika took 63 hours, or an average speed of 1 1/4 miles per hour. Descent 14 hours, an average speed of nearly 6 mph."

At the mouth of the Ingenika the men spent July 16 in camp while Swannell calculated his triangulations and Copley tied in the surrounding mountain peaks to the survey. The next day Swannell began plotting up his map of the Ingenika River. "Talking with Jim (the cook) about our Manson Creek stay a year ago, he tells of the old Klootch who much preferred the old wax vesta matches of 1871 – She and her child, when almost perished with cold, were saved by being able to start a fire with Vestas. 'Agham wake siyah mameloose; stick match delate cultus, glease match delate kloosh, hyak spitabee.'" On the 18th of July Swannell's crew left for Deserter's Canyon to set up a cache while Swannell remained in camp to continue working on his map. After the men returned on the 19th the survey crew moved camp across the Finlay River. From there Swannell and Copley backpacked down the river to Fort Grahame. Swannell wanted to pick up a few supplies and leave his notes and maps of the Ingenika River at the fort.

The men arrived at Fort Grahame on the morning of July 21. "Seven Sicannis are sick, five already dead. Disease unknown, probably largely starvation. Ross (the factor) not back, and altho' they can see food within the store they are too honest to break in. Are amazed that I dare break the lock and serve them out supplies. Reminded me of Butler in 'The Wild North Land' & the Indian Moose that Walks, 'Chap XXII' Yes; he had learned the lesson of honesty, but his Teacher, my friend, had been other than human." In a note added by Copley in 1955 he wrote: "I remember quite definitely that Jim

■ Big Bend of the Finlay River. Although the water is still swift the river is now shallow enough to pole in most places. Evidence of the forest fires that occurred when the Klondikers traveled through this area can be seen.

Nep Yuen & I collected wild strawberry leaves, boiled them up and concocted medicine with this infusion mixed with your maple syrup. This cured the Indians of their dysentery brought on by starvation and the boiled flour they were living on before you broke into the store." Copley also recalled that in 1913 at Fort Grahame: "Jim A. bought a pair of corduroy pants for $5 and after keeping them several weeks sold them to me for $1.00 because the pants had cuffs which he had never seen before. He thought they looked too sissified."

After returning from Fort Grahame, Swannell's crew continued up the Finlay River above the Ingenika. Two days of travel brought them to the foot of Deserter's Canyon. Samuel Black named this canyon because two of the men in his 1824 expedition deserted there. Swannell and his crew spent ten days at Deserter's Canyon. On the 24th of July they packed their "cache across the Portage Trail to the head of the canon. Very heavy rain from 4 pm on. Find tracks on the beach at the big eddy of two mountain goat which swam the river – one was drowned in the terrible whirlpool." The next day they portaged "our very heavy canoe, using a Spanish windlass in steep places – on top have to lay skids to drag the canoe over River, foot of canon has risen 3 ft. Fresh snow on the mountains." In "Ninety Years Later" Swannell observed that: "The Klondikers of '98 had laid skids across the portage, but these and their windlass were rotten and we had to renew them." On the 26th of July Swannell took photos and sketched the canyon. He observed that: "There is a bad cascade & whirlpool at the foot of the canon and no beach whatever up the canon. It would be impossible to line up even large bateaux. Trees uprooted above and floating down, some large spruce are sucked under at the head of the canon and never re-appear for several hundred yards." During their time at Deserter's Canyon Swannell recorded that the water above the canyon rose six feet in four days. Swannell surveyed Deserter's Canyon and continued his daily record of temperature, comparison of their watches with the chronometer, and measurements with the barometers.

From Deserter's Canyon Copley and Alexander took a trip to the west and discovered a string of small lakes that they named the Emerald Lakes. Swannell then went

■ Crew running Deserters Canyon. Despite his unhappiness at not running the canyon with the crew, Nep Yuen took this photograph of Swannell's crew.

with them on a two-day trip back to these lakes to survey and tie them to his Finlay River survey. On their second day they had misfortune with their dog, Dick. Patterson described this incident. "To guard the outfit they left dog Dick tied there through the whole of one long day. Loneliness beset the hound, the sun was hot, the steamy heat after the rains brought out an amazing crop of young and vicious mosquitoes – and, in his struggles to get loose, Dick upset his supply of water." Swannell recorded this misadventure in his journal. "Our big dog Dick, left tied up went crazy with the heat & flies and tore our mosquito tents up – hence a bad night for us." Copley added his recollection of this event. "Dog Dick not only tore up our tents, but also dug a hole and almost buried them up. When we came home and let him loose he left camp and went to the west side of the lake and stayed there all night. Several days later I found an old fly in an abandoned prospector's cache & made myself a lean-to which I used for the balance of the season."

On August 3 Swannell and his men left Deserter's Canyon. The next morning Swannell observed "snow down to timber on mountains." That day, as the crew surveyed up the Finlay River, Great Britain declared war on Germany. Little did the men realize that the reverberations from World War I would echo halfway around the world into the quietness of this remote valley where they were working, affecting their lives.

Above Deserter's Canyon the Finlay River was swift but there were no major obstacles for a while. The men continued surveying along the river with occasional visits to the tops of nearby mountains. Some days Swannell remained in camp calculating his river traverse while the men tended to various details. Beginning August 5 Swannell started noticing the leaves turning color. On August 7 Swannell noted that according to his calcula-

■ Lunch camp on the Finlay River. Nep Yuen has prepared lunch for the crew and laid it out on a blanket on the river bank.

tions the chronometer was slow by 2m 30.5 seconds. Then he reset the men's watches to the new correct time. Nep Yuen enjoyed fishing and was the best fisherman of the crew. One day Swannell wrote that: "The cook loses a 24" sapi [dolly varden] who carries away hook and gut cast. Two miles further up, on the opposite shore and over an hour later, he catches the identical fish, which must have followed the canoe. A 5 lb. fish."

By August 16 Swannell's crew reached Paul's Branch where they found "skeletons of two horses and their pack-saddles, one with 'Sousie' branded into it (Klondikers)." In "Ninety Years Later" Swannell wrote that: "The pack-trail of the Klondikers of '98 runs up the east bank; we found several cabins in which they had wintered. Pathetic records of their hardships were the skeletons of their pack-horses, rigging and cross-tree pack-saddles still in place as they died of exhaustion or were shot." Their next camp was about five miles up the river. "Main Liard Pack Trail along river here is very poor, not used since Klondike days. An old pack-train camp 1/4 mile down with two skeletons of horses – lying with their pack-saddles."

On August 21 Swannell's crew reached the "Kwadacha River (big white water) at noon. A swift deep muddy river laden with glacial silt, which does not mingle with the Finlay water for several miles below the junction. The Finlay above perfectly clear, getting swifter and shallower." Shortly after the Kwadacha the survey crew reached the Fox River where they camped for three nights. Up to this point Swannell's crew had traveled

■ Survey crew at Fort Grahame.

northwest following the Finlay River through the Rocky Mountain Trench. Now the Finlay turned to the west, out of the Trench. The Fox River continued up the Rocky Mountain Trench to Sifton Pass. Swannell probably hoped to travel up the Fox River. "George and Jimmy try to take canoe up Fox River but are stopped by a 3' overfall 1/2 mile up." Their next camp up the Finlay River was at a trapper's cabin at the foot of Prairie Mountain. On August 25 Swannell's survey crew "climbed Prairie Mt., steep smooth grass slopes and bluffs, flattish top. The Rocky Mountain Trench in which the Finlay has run bears N 30º W and is now occupied by Fox River. 4 trappers arrive in a big boat – French Canadians." These were the only people that Swannell's men met between late July and early October.

In this remote area of the upper Finlay Swannell's men could not replenish their supplies, so they tried to live off the land as much as possible. At the end of August Swannell made a record of the food that each man had provided. Nep Yuen had caught over 100 fish, while the other three men combined for 38. Nep Yuen had also picked thirteen quarts of black currants and cranberries. Copley had shot 100 rabbits, while the other three men had a total of sixty. Copley had also shot seventy-eight grouse. Jim Alexander and Frank Swannell had shot a variety of game including the big game animals, two moose and two goat.

By August 28 Swannell's crew reached Bower's Cache at the foot of Long Canyon, their next obstacle on the Finlay River. Before starting through the canyon Swannell spent a few days surveying up Bower Creek and climbing Bower Mountain. "Very cold and windy on top and impossible to read angles except in intervals of lull." While in camp Swannell learned "a little Stuarts Lake dialect from Jim."

On September 2 Swannell's crew started through

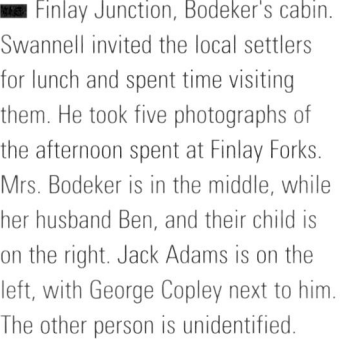

Finlay Junction, Bodeker's cabin. Swannell invited the local settlers for lunch and spent time visiting them. He took five photographs of the afternoon spent at Finlay Forks. Mrs. Bodeker is in the middle, while her husband Ben, and their child is on the right. Jack Adams is on the left, with George Copley next to him. The other person is unidentified.

Long Canyon. "We got thru all but the last half of the canon…Had not Jim Alexander been sick today we would have got to the head of the canon. We had to bivouac, each of us curled around a tree, as there was no level ground large enough to pitch a tent on." In his journal Swannell lists this site as Bivouac 67, instead of a camp. The next day they finished traveling through Long Canyon. Copley commented: "You might mention the quicksilver in a…bottle we found in a cabin near the head of Long Canyon. When we found the bottle we thought it was fastened to the floor somehow, it was so heavy. We used the mercury to try and clear out the lead in our 22 cal. rifle, but made the gun worse than ever."

From Long Canyon it took only two days to reach Cascade Canyon. Along the way Swannell passed "a large creek 60' wide by 9' deep…from the South which runs West paralleling the Finlay – probably the one ascended by McConnell, G.S.C. [Geological Survey of Canada] in 1895." McConnell was in this area in 1893, and the creek was Cutoff Creek that led to McConnell Pass.

It took Swannell's crew four days to get through the many sets of rapids in Cascade Canyon. The first day was the most difficult. "Worst place the cascade where four or five channels are cut through hard rock – Here we had to lay skids and haul up the canoe, 200 yards of bad water above but no chance to portage." On the last day the men carried their supplies to the head of the canyon and dragged up the empty canoe. "Some very dangerous water, including one cascade up which we poled and lined the empty canoe. The head of the canon lies between 100 ft. cliff."

Swannell spent three days at the head of Cascade Canyon. The first day Swannell gave the men a holiday while he spent time calculating his traverse. "Great excitement as dog Dick chases a moose into the river and dashes after it. We are afraid to shoot for fear of hitting Dick. Snowing hard in the mountains up river." The second day he "cruised across to near Fox River over the wide valley in continuation of the Upper Finlay Valley – This is the ancient bed of the river before it cut Cascade Canon."

For the next week Swannell's crew continued surveying up the Finlay River. One evening their campfire got "away in the night but we got it stopped at 3 am." On September 14th "Dick runs a moose and Jimmy shoots it from across the river and wounds it. It takes to the river and Jimmy drops his rifle & plunges in followed by our smaller dog [Caesar] – He climbs on top and cuts its throat. But guided by a long stick he has to ride it down a quarter mile before he can steer the moose onto the top of a bar – the little dog swimming alongside and nipping at the animal." They spent a day in camp cutting up the moose and smoking the meat. By the 17th of September they were at the Big Bend of the Finlay where the river turned south. "The river is now a succession of shallow boulder-strewn rapids 'like going upstairs'." In "Ninety Years Later" Swannell commented: "The valley bottom was worthless and stony, denuded of soil by the fierce forest fires which had swept over it – a legacy from the Klondikers."

Swannell's crew reached Reef Canyon on September 18. "Fall has set in with a vengeance!…Traverse

Running the Finlay Rapids on the Peace River. These are the rapids that Swannell, Alexander and Nep Yuen ran a second time when Copley forgot to take the picture.

upstream in the afternoon. Affairs look so bad that Copley and I scramble several miles up the right bank. The canons seem all but impossible and we nearly decide to abandon the water-route. However since a boy Jimmy has heard tell of the mysterious Fishing Lakes and we easily persuade him to tackle the canons." The men rested for a day before starting through Reef Canyon. In "Ninety Years Later" Swannell recalled that: "After forty years I well remember the little quiet cove and sandy beach between two jutting rock points where we camped. It is undoubtedly Black's 'L'Ence [Anse] du Sables' of 90 years before." Swannell's journal entry for September 19 noted that it had snowed "most of night and morning, later turning in to rain. Called today our official Sunday!"

Reef Canyon is two and a half miles long and it took the men three days to get through it. On the first day disaster struck at a place that Swannell called Kodak Cascade. The men emptied the canoe and portaged their outfit. However, the rock was "too steep to make it possible to make a road for the heavy canoe. Try to line it through a narrow chute between two rocks – the bow dips under and the canoe half fills. We had forgotten to empty the wee forecastle built in the bow to house chronometer, aneroids & chronometer [Swannell meant camera]. The chronometer box floats out and we save it undamaged; but the camera is water-soaked and the shutter ruined. (This is a major disaster and henceforth I can only take pictures by lifting on and off a wooden cap whittled to cover the lense.) We don't get far today." Fortunately for Swannell the rest of his film had been portaged. Swannell's pictures of Kodak Cascade were ruined. However, he dried and cleaned the camera so that it was functional again. Despite the fact that he had to make a wooden cap and manually judge the amount of light needed for each picture, Swannell was still able to take excellent photos during the rest of his 1914 field season.

At the head of Reef Canyon Alexander shot a caribou to add more meat to their food supply. The following day, September 22, the men reached the Fishing Lakes, "the mud beaches trampled like a barnyard by caribou" according to Swannell. Patterson wrote: "Whether the lakes came up to Alexander's expectations or not, I never heard." On the 25th of September Swannell's men reached the Thucatade River. "Below its mouth the trail to the Stickine crosses…Near the trail crossing we found good well-built clinker boat which had been cached about 6 years, also a small dug out canoe & a scribbled note 'Thanks awfully for the use of your boat – Pat Birch'."

The next day Swannell's crew made their last camp on the Finlay River about one mile above Delta Creek. Along the trail they found a tree with the name Denis Frank on it, and by their camp they found a mining claim post, dated 1909. At this camp they used the last of their oatmeal and sugar. Except for the meat they had recently acquired their food was running low. In "Ninety Years Later" Swannell explained their situation. "Our final camp was at Delta Creek, 14 miles above the canyons, where the river was only 40 yards wide. A scant

■ Hudson's Bay Company, Hudson's Hope. Swannell visited this HBC post when he was traveling through the Peace River area on his return to Vancouver.

thirty miles farther would have brought us to Thutade Lake, the head of the great Mackenzie River and some 2,500 miles from its mouth. But our food was now mainly dried moosemeat, the river was rapidly getting worse, and the weather cold and stormy, so we decided to abandon the survey and turn back."

The canyons were behind them. Swannell knew that he was close to Thutade Lake, but he didn't know the exact distance at the time. However, the weather was turning cold and there was snow on the nearby mountains. It was difficult getting the heavy dugout canoe up the shallow boulder-filled river. The water level of the river was dropping daily, which would make it more difficult to get down the Finlay. The Surveyor-General's letter had indicated that there was at least one more year of field work in the area, and Swannell probably thought that he would return in 1915 to complete the survey and map the lakes and headwaters of the Mackenzie River.

Swannell did not know that T.H. Taylor (PLS #51) had reached Thutade Lake on September 9. This was Taylor's third year surveying coal lots in the Groundhog area. Trygve Rognaas, who had worked for Swannell in 1910 and 1911, was Taylor's cartographer. Taylor left Hazelton on February 27. Traveling by dogsled and snowshoe he reached the 4th Yukon Telegraph Cabin on March 5. Throughout March and April he had dogsleds take supplies into the area where he would be surveying. Since he was in the area, Taylor was able to begin working as soon as the snow melted, and he was able to travel light since his supplies were already cached throughout the region. In mid-August Taylor made a quick traverse from the Groundhog region over the mountains and down to Thutade Lake where he left a conspicuous survey marker for Swannell. If Swannell had known that Taylor reached Thutade Lake he probably would have also made a quick traverse to the lake to connect their

■ Fort St. John, Peace River.

Teepee, Beaver Indians, Fort St. John. While he was at Fort St. John, Swannell visited a group of Beaver Indians who had come to the community. These First Nations people used tipis for their lodging like their neighbouring tribes on the Prairies.

surveys. This would have provided a major, historic surveying link across northern British Columbia. The delay in reaching Fort St. James in the spring had major consequences for Swannell, for another week of good weather would have enabled him to reach Thutade Lake.

The men stayed at Delta Creek on September 27. "George and Jimmy hew out a la Robinson Crusoe two board, each from a tree trunk. These we nail on the canoe with a flare outboard so as to have more freeboard and save us from swamping in the canons below." About this invention, Patterson wrote: "Besides providing more freeboard, these flaring extensions of the gunwales would tend to fling the broken water in the canyons outwards and away from the canoe. That was a thing of Alexander's devising, and it undoubtedly saved them from swamping and drowning in the wild water they had to run." Nep Yuen cooked and fished while Swannell continued his surveying. "I cruise, track-traverse & sketch six miles up the trail following the river to where the trail goes up a 30 ft. creek. This creek is the fork of a 60' torrent entering the river thru a swamp with several beaver houses. At the trail crossing of the 30' creek are 4 posts crowned with stone caps. One and a half miles above our camp is a Stickine Siwash grave with a flagpole and rifle." R.M. Patterson described the scene: "On the way back he stopped to look again at the grave of a

Stikine Indian that was close by the trail. A flagpole had been raised over it. From its peak a wind-torn wisp of some material still dangled limply in the sunshine, motionless on this quiet autumn day, and to the base of the pole was secured the dead man's rifle. Swannell would be a man of over fifty before he saw that grave again, no longer young but still with the vigour of youth, and with exciting years behind him." As he returned to camp, Swannell could not have imagined that it would be seventeen years before he would return to this place again.

On September 28 Swannell's men began their descent of the Finlay River. "Rain and snow nearly all day and night and exceedingly miserable in the canoe." The next day the men ran Reef Canyon. R. M Patterson described their adventure in this white water.

Copley writes:

"We ran this canyon at the peril of our lives – we were very short of grub and didn't want to take the time to portage, which would have delayed us two days. We looked the water over carefully, then walked back to the upper end and went into conference – and finally decided to take a chance. If we had upset we would all have drowned as no one could possibly have lived in that turbulent water.

What happened after the conference was that Jim Alexander took each one of them aside and asked separately if he would take a chance with him. He got hold of Copley first – and he agreed. Then he asked Swannell and he said yes. Nep Yuen also agreed, and he gave his reasons which ran more or less like this: "Better I go with you. No good watch you fellows drown and then starve and freeze to death in the bush. Better I drown, too. So I go." This frank statement cheered everybody up and was known from then on as 'the logical remark.'

Swannell described their trip through Reef Canyon.

Reef Canon very bad and Kodak Cascade worse.

A big wave came up intermittently in the eddy at the foot of the cascade – we ran ashore above the cascade to size up the situation and hoped to miss the wave – However we miscued and the wave rose – I in the bow, had it break over my head, it was 6' high. The smother went clear over the canoe. We shipped six inches of water and had to run ashore in a hurry.

Worst water I ever saw but we had no option. It had to be run, lining or dropping down with the pole being impossible – We should have portaged the outfit…

In a note included later, Swannell added more detail.

At a bad turn in Reef Canon an acute turn had to be made directly in face of a cliff into which the current impinges violently. Jimmy didn't think he could make it and shouted to me to stop the canoe with the pole by tilting at the cliff. The impact threw me overboard but George grabbed me and hauled me on board.

Copley also remembered that: "When we swept down Jim Nep lay down in the bottom of the canoe and got properly wet." In "Ninety Years Later" Swannell used a photograph that he took of Reef Canyon to illustrate the route they used to run the canyon.

The following day, September 30, Swannell's crew went through Cascade Canon. "Dropped down the bad water to the top of the main cascade on the line and nearly swamped. Strike rocks twice in the rapids below. George is dragged into the river, while ashore lining, but Jimmy drags him out, reeling him in like a fish." Copley noted that his clothes were frozen solid by the time they made camp. He also recalled that "when I was thrown into the river at this point my big nickel-cased Waltham watch filled with water, but being dried out at the camp-fire later on started to run and kept as good time as ever."

By October 1 Swannell's crew was at Long Canyon.

Get through Long Canon without incident. At the foot of Long Canon we run ashore to investigate a red flag hung out on a sweeper to some trapper who knew we were above. It had wrapped in canvas an August newspaper with huge headlines – The Declaration of War. Our last contact with a human being had been the French Canadians at Prairie Mt. – but they had come from the Liard and evidently never heard of the war.

At Bower Creek meet Hedges (After I had gone to France Hedges enlisted in Victoria in 1915 and called on my wife, asked her to remind me of the trapper with the silver-fox pelt he met on the Finlay)…Met Brandon & Warner at Fox River, Booth & Bennett at the Kwadacha & Isaac Robinson reached here on a raft…But for the newspaper we wouldn't have credited their stories of the war.

Two days later Swannell's crew was back to Deserter's Canyon where they had been in July. "Run through with the whole load excepting the blanket rolls. Nep Yuen is very angry at me because I put him ashore at the head and so he misses the fun of running through…The water at our old high water landing at the head of the canon has dropped 8'."

On October 4, seven days after leaving from the upper Finlay River, Swannell's crew arrived in Fort Grahame. Swannell wrote: "To Jimmy belongs all credit

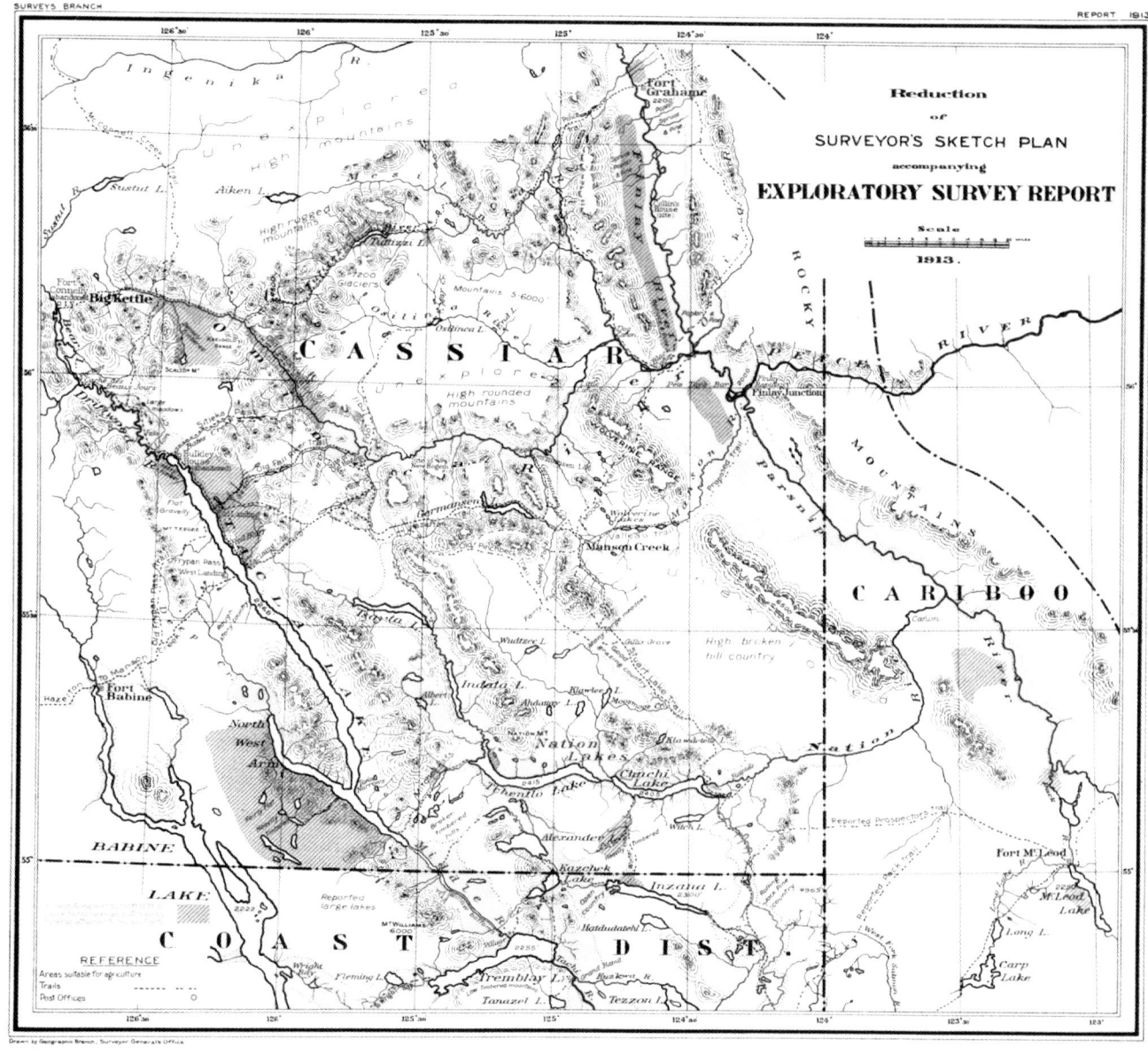

■ This is a copy of the large, detailed map (about five feet by three feet) that was made based on Swannell's exploratory surveys of 1913 and 1914. The map was not completed until 1917 because of World War I. This map is well known for it is the first map produced by surveying that shows the geographical features of a large part of northern BC.

for getting us safely down the Finlay." Swannell took observations on Polaris. He also spent time examining the Hudson's Bay Company journal and copied notes about some of the Klondikers who passed through Fort Grahame. "Jim Alexander tells me he brought letters on snowshoes for these men from Fort St. James and how disgusted one man was who got, and paid for, 10 letters and found them all bills." Swannell spent four days at Fort Grahame surveying the Hudson's Bay Company Land and an Indian reserve. On the way down the Finlay River Swannell noticed several fires burning on the hillsides. By October 10 the crew were at Finlay Forks, and Swannell invited several of the local settlers, whom he had first met last fall, for lunch. "Run Finlay Rapids purely to obtain photos, but Copley stationed on a boulder as photographer is so entranced at our speed that he forgets to trip the shutter. So we line back up and repeat." Swannell's famous trip on the Finlay River was finished. His description of the land and his photographs are especially meaningful for Finlay Forks, Fort Grahame and the lower portion of the Finlay River are now submerged under Williston Lake.

Swannell decided to return by a new route this fall, so at Finlay Forks Jim Alexander left to return to Fort St. James. On October 12 Swannell, Copley and Nep Yuen started down the Peace River in their dugout. Swannell recorded that this was Day 163 and Camp 96. The next day they went through the Parle Pas Rapids which were easy to navigate at this time of year. By October 14 they arrived "at Rocky Mt. Portage head of the Canon 4:30 pm, camping at Cust House 1/4 mile above the head of the Canon. The next day they walked "across Portage Trail to Hudson Hope and arrange with Del Miller, a young chap with a shy pretty Indian wife to pack canoe over for $25 and outfit at 2 1/2 cts. a lb." On Friday, October 16 they returned back to the other side of the portage, and their packers arrived in the afternoon. The next day the packers took across their "canoe and outfit substituting a lighter abandoned canoe for our own and turning our own loose into the canon." Swannell and his men spent the 18th of October at Hudson's Hope where they visited the Hudson's Bay Company manager. They also walked "up to the foot of the Canon and find a broken piece of the gunwale of our old canoe…Visit Beaver Indian camp housed in skin tepees."

After leaving Hudson's Hope, two days of leisurely canoeing brought Swannell, Copley and Nep Yuen to Fort St. John. "The manager Beaton is down river, but his Indian wife and half-breed boys are home and provide us with a cabin. G.B. Milligan BCLS and his assistant Cartwright are here. He left Victoria May 1913 and came north via Athabasca Landing. Has been exploring the NE corner of BC lying north of the Dominion Peace River Block examining 20,000 square miles. He wintered at Fort Nelson. We arrange to go out together." On October 23 Swannell's crew left Fort St. John, continuing down the Peace River. En route they met "Beaton & sons & "Reveillon's Manager tracking upstream." The next day they crossed "the newly run Alberta Boundary at noon." The following day they camped "about 70 miles below Fort St. John at a big loop above a wrecked side-wheel steamer. Get nearly stuck in a slough. Lunch with Trapper Hubble."

On October 26 they arrived at Dunvegan, where Swannell and his men waited for Milligan, who didn't arrive until the following day. By the 29th of October the two groups of surveyors arrived at Peace River Crossing where Swannell sold his canoe. Copley recalled an amusing incident at Peace River Crossing.

> Don't forget the episode of Cartwright and his gal jug of rum at Peace River X ing. From what I remember he ordered the rum two years before but it arrived in Peace River X ing after they had left to go north. He told us all about it at Dunvegan and how we would enjoy it on our arrival at P.R. X ing. Shortly after we got there we went to the store and sure enough it was there OK with the HBC seal unbroken, so we took it up to his room, got hot water, sugar & glasses Broke the seal and opened the jug. What a surprise the jug was filled with water. It had been tapped on the way from Edmonton some way or other.

Swannell and Milligan's survey crews left Peace River

Crossing by wagon. After two days of travel they arrived at Lesser Slave Lake where they boarded the *Northland Sun* which took them to Sawridge. Then they drove to the rail line where they boarded a train that took them to Edmonton, arriving on November 3.

Swannell spent two days in Edmonton. While he was there he invited "to supper Charlie Hemeyer and his sisters. He an old Nechako survey hand." The next day Swannell was "taken all over the university by Smith – Miss Hemeyer's fiancé and to the theatre to see 'Peg o' my Heart'." On November 6 Swannell left by train for Calgary, and the following day departed for Vancouver. "Armed guards at every bridge and tunnel. Am seen taking a photo for Castle Mt. Am suspected of being a German spy and questioned by a self-constituted committee of passengers – a beautiful instance of stupid hysteria."

The surveyors arrived in Vancouver on Sunday, November 8. Copley had a long discussion with John Davidson, the Provincial Botanist, about the plants that he had collected in the Finlay basin. Swannell spent the next day in Vancouver. He met Victor Kastberg, the forester who had been with his crew in 1913. Swannell also saw T.H. Taylor and discussed the Groundhog surveys, and visited McElhanney to talk about his work on the 124th Meridian. Copley conferred with the Provincial Botanist again. That night Swannell, Copley and Nep Yuen took the boat to Victoria, arriving there on the morning of November 10.

At the end of December Swannell wrote a report to the Surveyor-General on his exploratory survey of the Finlay and Ingenika valleys. Most of his report is a description of the Finlay and Ingenika Rivers and their tributaries. "In accordance with your instructions, the reconnaissance was executed primarily with a view to the production of a topographical map on a scale of four miles to the inch. To attain this end a rough triangulation was carried over the area cruised; the Finlay River was traversed for some 125 miles and about 160 miles of track traverses were run." Swannell wrote that: "The area is approximately 5,000 square miles. The main characteristic of the whole region is its mountainous character, only 400 square miles being agricultural lands…The total population of this 5,000-mile area consists of a scant dozen of white prospectors and a couple of small bands of the nomadic Sikanni Indians. No attempts at agriculture have been made, except a little desultory gardening at Fort Grahame. The Finlay Valley, however, as far north as 57º 30' contains probably the largest compact area of excellent land remaining unexploited in British Columbia, and both soil and climate render it particularly desirable for settlement. Swannell added a note of caution. "At present remoteness, and consequent difficulty and expense of obtaining supplies, is the chief drawback." In his report Swannell mentioned that: "A botanical collection of some 1,500 specimens, comprising thirty plant-families and forty-four varieties of seed, was made by my assistant, Geo. V. Copley."

Swannell also produced a separate summary of his 1914 survey work. He included the dogs as part of his survey party. Swannell noted that he surveyed from 453 stations, taking a total of 5,700 transit readings. The longest sight between stations was 115 miles. Of the 5,700 transit readings 700 were for astronomical observations, and 1,000 were vertical angles for altitude. Swannell took 600 barometric readings and 560 thermometer readings. There were 33 astronomical readings for latitude, 87 for time and longitude, and 20 for azimuth (direction). His itinerary for the season included 3,680 miles of travel, including 1300 by railway, 1200 by canoe and 800 by steamboat. Swannell listed the game that they procured. There were 134 fish (mostly caught by Nep Yuen), 203 rabbits (mainly obtained by Copley), and 166 grouse. Swannell and Alexander got the five big game animals (two moose, two goat, and one caribou). Swannell also noted that the Provincial Botanist reported that Copley had collected some rare specimens.

World War I virtually ended government surveying in British Columbia until the end of the war. Although there was much work to be done on the field notes from 1914, Swannell was already planning to enlist in World War I. It would be six years before Swannell returned to northern British Columbia, and his life, and the way of life in that region would change dramatically during that time.

Summer 2004—90 years later

It's been 90 years since Frank Swannell finished his first period of surveying in northern B.C. The region has changed considerably during these years.

The Nechako Valley is completely settled, and is the main agricultural area for that part of the province. South of Vanderhoof the upper Nechako has been extensively logged. Kenney Dam and the Nechako Reservoir have flooded the Cheslatta Reserves and many of the lakes and waterways of the Great Circle tour that Swannell took in 1910. The water level of the Nechako River is considerably lower most of the year.

The Hudson's Bay Company post at Fort St. James is a National Historic Park, and most of the buildings that were present in Swannell's time still remain, including the schoolhouse where he usually slept. Mount Pope, where Swannell began his exploratory surveys in 1912 and 1913, is a provincial park, and provides an excellent view of Stuart Lake and surrounding mountains.

The Stuart River and the Middle River, between Trembleur and Takla Lake, have been designated provincial heritage rivers.

North and west of Fort St. James logging and logging roads have divided the once-extensive wilderness of northern B.C. into smaller sections. However, pockets of wilderness remain. Omineca Provincial Park and Sustut Provincial Park protect sizeable portions of the Omineca Mountains. Mount Blanchet, at the junction of Takla Lake and the north arm of Takla Lake is a provincial park. Tsayta and Indata, the upper two Nation Lakes, have also received provincial park status.

The lower Finlay River is now covered by the Williston Reservoir, and Fort Grahame and Finlay Forks now lie under water. Above the Fox River a large

■ Looking down the Fraser River from the sternwheeler landing in Quesnel

■ Takla Lake from Lovell Cove

■ Kastberg Creek. Swannell followed the Ingenika Trail up this creek into the Omineca Mountains in 1913.

■ This is the schoolhouse/ men's house where Swannell stayed when he was at the HBC post in Fort St. James. Swannell took a photograph at this same location when he was preparing to depart on his 1914 exploratory survey.

■ A view of Stuart Lake and the surrounding mountains from the summit of Mt. Pope, where Swannell started his exploratory surveys in 1912 and 1913.

section of the Finlay is protected by the Finlay-Russell Provincial Park.

This summer, thanks to BC Rail, I was able to ride on their UTV that travels between Lovell Cove on Takla Lake, and the Canfor logging camp at Minaret. The upper portion of the Driftwood River and Bear Lake are still wilderness except for the rail line. The old Native village at Bear Lake has no permanent residents. During the day that I spent there I had a sense of the history of the site. I was able to find some of the places where Swannell took his photos when he was surveying the reserves in 1911. I could look down the lake from the boat launch beach and gaze at the same scene that many well-known Native and non-Native people had seen. Along the upper Driftwood River I could see the piles of driftwood for which the river is named, and view the waterway that Swannell traveled in 1911 and 1913 still undisturbed. At other places I could catch a glimpse of the northern BC that Swannell saw between 1908 and 1914. From the top of Mt. Pope I could see many of the mountains that Swannell surveyed. The schoolhouse at Fort St. James National Historic Park appears ready for surveyors and miners to stay there again. At Quesnel I stood at the landing where the sternwheelers docked, and looked down the Fraser River that appeared largely the same as 90 years ago. A similar scene greeted me at Soda Creek, where Swannell boarded the sternwheelers.

I would like to thank all the people who were so hospitable and friendly during that week in northern BC. Your enthusiasm for this book and your assistance is greatly appreciated.

Appendix I

The Men of Swannell's Surveys

Many men worked on Swannell's survey crews from 1908 to 1914. Often these people were from the area where Swannell was surveying. Most of them worked for one season or less. Some men, who were learning surveying, or who liked the rugged outdoor life of a surveying crew, worked longer. George Copley, Swannell's close friend, surveyed with Swannell for six seasons. Patrick Sharkey, the cook, worked for five seasons, while Vilhelm Schjelderup, Jim Alexander, Nep Yuen and Lawrence Dickinson worked for three. Swannell kept in touch with many of his survey crew workers throughout his lifetime.

Swannell made a list of the men on his crew each year, but this does not necessarily include all the people who worked for him.

1908
Abbott, Edgar
Benson, R. M.
Bishop, R. P.
Davidson
Deschamps, Bellamy
Deschamps, Henry
Driver, Henry
Forbes, Gilbert
Hemeyer, Charles
Hills, Dad
Indian Daniel
Indian David
Indian Johnny
Indian Michel
Macdougall, Archie
Macfie, John
Robertson, A. I.
Sackrider, H.
Sharkey, Patrick - Cook
Sturgess, A.
Swinerton
Thorpe, Joe
Wilkie, W.
Wilkins, G.
Wyllie

1909
Bartlett
Benson, R.M.
Copley, George
Goldie, Jack
Goldie, Hugh
Greer, Tom
Hemeyer, Charles
Hill, John
O'Meara, Alf
Pickard, George
Schamerhorn, L.H.
Schjelderup, Vilhelm
Sharkey, Patrick - Cook
Thorpe, Joe.

1910
Braithwaite, John
Copley, George
Currie, John
Danielson, Arne
Dickinson, H.W.
Dickinson, Lawrence
Ellis, Sam
Greer, Tom
Hallam, William
Hemeyer, Charles
Hetherington
Ketlo, William
Larson, Olaf
MacDonald, Pete
McDonnell, Pat
McMeekin, Jim
O'Meara, Alf
Olson, Anton
Pike, Francis
Rognaas, Trygve
Sackrider, H.
Schamerhorn, L.H.
Schjelderup, Vilhelm
Sharkey, Patrick – Cook
Spetch, Sid
Thompson, Norman
Thompson, Roy
Waters, W.E.
White, George

1911
Bishop, R.P.
Copley, George
Currie, John
Danielson, Arne
Dickinson, Lawrence
Hallam, William
Hemeyer, Charles
Holland, Alwin
Ketlo, Ashiel
MacDonald, Pete
MacDougall, Douglas
Prince, Albert
Prince, Jean Marie
Prince, Maryaz
Rognaas, Trygve
Schjelderup, Vilhelm
Sharkey, Patrick - Cook
Thompson, Norman
Thompson, Roy
Thorpe, Joe
Welch, Percy
White, George

1912
Alexander, Jim
Copley, George
Currie, John
Dickinson, Lawrence
Holland, Alwin
MacDonald, Pete
MacDougall, Douglas
Mitchell, J.B – Forestry
Nep Yuen - Cook
Sharkey, Patrick – Cook

1913
Alexander, Jim
Copley, George
Gold, Axel –Forestry
Kastberg, Viktor – Forestry
Nep Yuen - Cook
Rossette, Sam

1914
Alexander, Jim
Copley, George
Nep Yuen - Cook

Several of the men on Swannell's survey crews have geographical places officially named for them.

Mount Swannell, in the upper Nechako River area, is named for Frank Swannell, along with the Swannell River, a tributary of the Ingenika River, and the Swannell Ranges in northern BC.

Copley Lake, south of Vanderhoof, and Mount Copley, northwest of Fort St. James are named for Swannell's assistant. In a letter to the Committee on Geographic Names that Copley wrote when he was 95 years old he recalled climbing Mount Copley "with two men and erected a ∆ [triangulation] station, reading shots to adjacent hills and also to Mt. Blanchet in the forks of Takla Lake." Copley also remembered how Copley Lake was named after him.

> We were running short of supplies and yet had a considerable amount of work to do in the area. Accordingly, Mr. Swannell with one man dropped down the river to Fort Fraser HBC store at Fraser Lake to purchase and bring in more supplies via the old Hallett Lake trail with pack horses. Before leaving camp he gave me instructions to cut a trail through from our camp to connect up with the Hallett Lake trail at the western end of Hallett Lake. At that time this was unknown country to us. I was delayed a few days with other work, then I took two men and got started on the trail and we ran directly into the centre of this small lake about one mile in length. Of course we didn't know the size of the lake and decided to back up and cut the trail around the west end. This mishap delayed us about one day, then as we proceeded further we ran into other obstacles with more delays. In the meantime Mr. Swannell with one man and a small pack train had arrived at the west end of Hallett Lake, and spent a day *cussing* & firing an occasional shot, to give us some direction as to his whereabouts. On the fourth day we got through and when the story got into camp, the little lake was known afterward as "Copley Lake."

Nep Yuen is the only person of Chinese origin to have four geographical places in the province named for him. Yuen Creek, Yuen Lake, and Nep Peak, (overlooking Yuen Lake) are in the Omineca Mountains where Frank Swannell and his crew surveyed in 1913. Mount Yuen is in the Finlay River area.

Mount Alexander, south of the Nation Lakes, is named in honour of Jim Alexander.

Rognaas Lake, in the upper Nechako River area, recognizes Trygve Rognaas' survey work in this area in 1910 and 1911. Mount Rognaas and Rognaas Creek, along with Mount Trygve, Trygve Lake, and Trygve Creek are located in the headwaters of the Finlay River and honour Rognaas' later work in that area

Schjelderup Creek in the upper Nechako River area is named for Vilhelm Schjelderup, who surveyed with Swannell there in 1910 and 1911.

Mount Greer and Greer Creek in the upper Nechako River area are named for Tom Greer, who was a member of Swannell's crew in 1909 and 1910.

Lawrence Lake, in the upper Nechako River area, is named for Lawrence Dickinson, who worked on the surveys in this region in 1910 and 1911.

Norman Lake is named for Norman Thompson, who worked on the surveys in the upper Nechako River area in 1910 and 1911. Norman and Lawrence Lake are only a few miles apart.

Goldie Creek in the Nechako Valley is named for the Goldie brothers, Jack and Hugh.

MacDougall Lake and MacDougall Creek are named for Douglas MacDougall and are located near North Takla Lake, in the area where he worked with Swannell in 1912.

Mitchell Pass and the Mitchell Range, between Takla Lake and the Nation Lakes, are named for Jack Mitchell, the Forest Service representative who accompanied Swannell in 1912.

The two Forest Service men who accompanied Swannell in 1913 have places in the Omineca Mountains that recognize them. Axel Gold is remembered in Axelgold Peak and the Axelgold Range, while Kastberg Creek is named for Viktor Kastberg. Kastberg also worked for Swannell after World War I in the Tweedsmuir Park area and Kastberg Mountain, which he climbed, is named for him.

Mount Rosetti, south of Fort St. James, is named for Sam Rossette.

Albert Lake, in the Nations Lake area, is named for Albert Prince, while Jean Marie Creek, in the same drainage, is named for Jean Marie Prince.

Appendix II

List of British Columbia Archival Photographs

I-67729	*SS Nechacco* in Standing Rock Canyon (cover photo)
I-58116	Running transit at Lac La Hache
I-58165	Frank Swannell at Takla Lake
I-33700	*SS Charlotte*
I-33701	Survey camp opposite Quesnel
G-03735	Indian burial site at Graveyard Lake
I-58460	Making flapjacks for the survey crew
I-59727	Tsinkut Falls
G-03736	Indian axemen
I-67741	Swannell survey: Muskeg, no bottom with 9 foot picket
G-03734	Indian girls at Nulki Lake
G-03880	Stoney Creek Indian Church
I-67740	Daniel and his family at Stoney Creek Rancherie
I-58104	Jenny Blench and her father, Tom
G-07798	Swannell's surveying party at Fraser Lake
I-58465	Surveyors taking sights at Eyrie triangulation station
I-59875	Fraser Lake
G-06441	Vital's ferry, Cataline's train
I-58229	Travelling by scow from Fraser Lake to Soda Creek
I-57574	Sliding down a snowslide on makeshift sleds
I-33280	Landing in Fort George Canyon
I-33267	*SS Nechacco* being repaired, Nechako River
I-57406	Ludwig's tent
I-33256	Sharkey in camp on the Nechako River
I-57403	Hoy & Johnson's cabin
I-57415	Cleaning salmon at Stuart Lake
I-33184	Cleaning salmon - Stuart Lake
G-03741	Indians smoking salmon heads at Stuart Lake
I-33272	A Hudson Bay boat at Fort St. James
I-33273	Landing freight at Fort St. James
I-57388	Takla Lake 1909
I-33259	Surveyors at the mouth of the Tache River
I-57412	HB boat, Siwash up mast at Fort St. James
I-57408	Coccola and deer, Stuart River
G-03742	Sophie and Alice at Nautley Rancherie
G-03743	Salmon weir at Fraser Lake
I-33277	Cataline's pack train ñ near Burns Lake cabin
I-57422	Yukon Telegraph Trail, Bulkley Valley
I-33287	*SS Distributor* near Kitwanga, Skeena River
I-57433	On the Skeena
G-06450	Waterfront, Prince Rupert
I-33290	Prince Rupert 1909
I-57546	Fording the Upper Lillooet River
I-33599	Camp at Big Point ñ Upper Lillooet
I-67807	Frank Swannell with two pair of snowshoes
I-67808	Group on Seaton Lake
I-58407	Freight team, Cariboo Road
I-33671	Freight wagons and automobiles at 153 Mile House
I-57986	Central Fort George
I-33601	Blairís Store, Central Fort George
I-33918	*SS Chilco* at Giscombe Portage on the Fraser River
A-04057	The *SS Chilco* at Isle de Pierre Canon, Nechako River
I-58177	*SS Chilco*, Upper Nechako River
I-58162	Wooding up ñ Upper Nechako River

I-59617	Grand Canyon of the Upper Nechako River	I-57868	Sternwheeler *SS B.X.* in Fort George Canyon
I-33174	Chief Louis & family, Ootsa Lake	G-03877	William Ketlo and family, Nechako Road
I-33175	Whitesail Lake	I-33198	Indians playing lazy stick
I-59815	Lagoon on Whitesail Lake	D-03694	Dominion Day celebration, Ft. St. James 1912
I-67805	Swannell's survey crew, upper Nechako River	D-06395	Fort St. James Indians at Dominion Day celebrations
I-67742	Mr. and Mrs. William Bunting	I-59768	West Takla Lake
I-33909	Surveyors on lower canyon of upper Nechako River	I-58204	Triangulation Station, Mt. Blanchet
I-67739	Frank Swannell at transit in snow	I-58195	Frank Swannell at Takla Lake, reading triangulation
I-58387	Survey party on the Nechako Road using Constantineau's sleigh	I-67802	Discussing the route to the Nation Lakes
I-59662	Ferrying horses across the Fraser River at Quesnel	G-06447	Albert's Camp, Upper Nation Lake
		I-58203	Camp at Lower Nation Lake
I-58117	En route to Quesnel on the BC Express Company special	H-04827	Fort St. James
		I-58201	Jim Yuen melting snow for water
I-33914	*SS Fort Fraser*, Sestino Rapids, Nechako River	I-58192	Lunch, Deep Creek Cabin, Nechako Road
		I-58384	Larson's team outside the Occidental Hotel
I-33186	Donald Tod family, Fort St. James	I-33666	Surveyors at 105 Mile House
I-58221	Greer Valley survey crew performing gymnastics	A-03207	Surveyors on the boiler of the wreck of the *SS Enterprise*
I-58181	"The Scientific Guys"	I-67727	Kispiox Margaret and baby
I-33127	Mrs. A.C. Murray and daughter	I-33190	Kispiox Margaret and baby
I-33172	Indian rifle team at Fort St. James	I-33191	Plug Hat Tom and canoe crew
I-33173	Indian boy dancing	I-33192	Bear Lake Tom and family
SW-03819	The winning team	I-58213	Babine Trail, McPherson and Daniel Teegee
I-58186	Sports at Fort St. James; Pete jumping	I-67735	Babine, HBCo 1913
I-33166	Tremblai Lake Joe	I-33541	Jim Alexander at the site of Bulkley House
G-06445	Jean Marie and Matyaz making canoe poles	I-59795	Ta-wat-enslinstay Rapids, Driftwood River
I-33958	Survey crew poling lashed canoes on Driftwood River	I-58373	Pack train horses grazing in a meadow along the Ingenika Trail
I-59810	Portage at Cache des Beaux Jours into Bear Lake	I-59919	Driftwood glaciers
		I-59913	Pulpit Glacier from Driftwood-Omineca Trail
I-59825	Bear Lake, site of old Fort Connelly		
I-33165	Departure of Father Coccola from Fort Connelly	I-58002	"Big Kettle," Omineca River
		I-59952	Notch Peak Station
I-33167	Indian women at Fort Connelly	I-58380	Packing to Too-Tizzi Lake
G-06641	Sicanni Chiefs at Fort Connelly	I-59800	Moving camp down the Stranger River
I-59769	Fair wind and lashed canoes, Takla Lake	H-03708	Swannell Party at Fort Grahame
G-06443	On the Cheslatta Trail	I-57317	Warehouse at Fort Grahame
G-06637	The Bella Coola Trail	F-08608	Survey Crew Making a Canoe Near Fort Grahame
G-06444	Chief Louis's Daughters		
I-33169	Cheslattas at Stilachula		

G-06446	Indian Graveyard, Fort Grahame
I-59748	Lining up Black Canyon on the Omineca River
I-33994	Summit Moody Trail, Wolverine Range, Omineca
I-58015	Manson Creek, Omineca
I-58199	Ah Hoo, a miner of 1871
G-06434	GTP construction at Decker Lake
I-58156	Leaving Fort St. James for McLeod
I-58057	Indian Village, Fort McLeod
I-33187	Fort McLeod, meridian altitude for latitude
I-57298	Working up the Finlay at high water
I-33213	Tracking up the Finlay River
I-33215	Taking out the Finlay River fur, HBCo boat
I-33182	Sikanni chief Charlie Hunter, Fort Grahame
I-33224	Mooseskin boat, Fort Grahame
I-58166	Sunday work, Ingenika River
F-08609	Crew at Windlass Portage in Deserters Canyon
I-33206	G.V. Copley at lobstick signal on Paul Mountain
I-67728	Cascade Canyon
I-33232	Survey crew drying moosemeat at Big Bend on Finlay River
I-57326	Big Bend of the Finlay River
I-57328	Foot of Fishing Lakes on the Upper Finlay River
I-33205	Crew running Deserters Canyon
I-58174	Lunch camp on the Finlay River
I-33222	Survey crew at Fort Grahame
I-58169	Finlay Junction, Bodeker's cabin
I-59806	Running the Finlay Rapids on the Peace River
I-67803	Hudson's Bay Company, Hudson's Hope
I-33963	Fort St. John, Peace River
I-33180	Teepee of Beaver Indians in Fort St. John

Sources Consulted

Books

Bond, Courtney. *Surveyors of Canada, 1869-1967.* Ottawa: Canadian Institute of Surveying, c1966.

Butler, William Francis. *The Wild North Land.* Edmonton: Hurtig, c1968.

Deeper Roots and Greener Valleys. Compiled by Fraser Lake & District Historical Society. Fraser Lake: Fraser Lake & District Historical Society, c1986.

Harvey, R.G. *Carving the Western Path: by River, Rail and Road through Central and Northern B.C.* Surrey: Heritage House, c1999.

MacGregor, James G. *Vision of an Ordered Land: the story of the Dominion land survey.* Saskatoon: Western Prairie Producer Books, c1981

Morice, Father A.G. *The History of the Northern Interior of British Columbia.* London: John Lane, 1906.

Mulhall, David. *Will to Power: the missionary career of Father Morice.* Vancouver: University of British Columbia Press, c1986.

Neering, Rosemary. *Continental Dash: the Russian-American telegraph.* Ganges, B.C.: Horsdah & Schubart, c1989.

Ormsby, Margaret. *British Columbia: a history.* Toronto: Macmillan of Canada, c1958.

Patterson, R.M. *Finlay's River.* Toronto: Macmillan of Canada, c1968.

Peace River Chronicles. Selected and edited by Gordon Bowes. Vancouver: Prescott Publishing Company, c1963.

They Call Me Father: memoirs of Father Nicolas Coccola. Edited by Margaret Whitehead. Vancouver: University of British Columbia Press, c1988.

Thomson, Don W. *Men and Meridians: the history of surveying and mapping in Canada.* Volume 2. Ottawa: Queen's Printer, 1967.

Vanderhoof: "The town that wouldn't wait. Compiled by the Nechako Valley Historical Society. Vanderhoof: Nechako Valley Historical Society, c1979.

West, Willis. *Stagecoach and Sternwheel Days in the Cariboo and Central B.C.* Surrey: Heritage House, c1985.

Whittaker, John A. (compiler and editor). *Early land surveyors of British Columbia (P.L.S. Group).* Victoria: Corporation of Land Surveyors of the Province of British Columbia, c1990.

Woodcock, George. *British Columbia: a history of the province.* Vancouver: Douglas & McIntyre, c1990.

Magazines, Booklets, Pamphlets, Newspapers

Andrews, Gerry Smedley. "Frank Cyril Swannell, 1880-1969: pioneer British Columbia Land Surveyor" – transcript of speech for the B.C. Provincial Museum, June 20, 1979

Andrews, Gerry Smedley. "Surveys & Mapping in British Columbia Resources Development". 7th British Columbia Natural Resources Conference, Victoria, February, 1954. Transcript of speech.

Cariboo Observer (Quesnel) – newspaper – 1909 to 1913

Copley, George V. "Surveying Central B.C. in 1910". *B.C. Outdoors,* May-June, 1968, pp. 56-61

"Frank Cyril Swannell, B.C.L.S., D.L.S. 1880-1969". *B.C.L.S. Proceedings – 1970., pp. 117-119.* Victoria: Corporation of Land Surveyors of the Province of British Columbia. Report of proceedings of the sixty-fifth annual general meeting.

"Frank Cyril Swannell". *The Canadian Surveyor,* March, 1970. Volume 24, No. 1, pp. 164-165.

Hooper, Craig. "Sternwheeler days on the Nechako" Vanderhoof, *Omineca Express-Bugle, June 22, 1983, pp. 14-15*

Professional Land Surveyors of British Columbia. "Cumulative Nominal Roll, 4th[th] Edition. Victoria Professional Land Surveyors of British Columbia, 1978

Professional Land Surveyors of British Columbia. "Cumulative Nominal Roll, 8th Edition. Victoria: Professional Land Surveyors of British Columbia, 2002.

Provincial Archives of British Columbia. "Frank Swannell: British Columbia Land Surveyor" Pamphlet produced for photo exhibit, 1979

Swannell, F.C. "Ninety Years Later". *The Beaver,* Spring, 1956, pp. 32-37

Taylor, W.A. *Crown lands: a history of survey systems.* Victoria: Crown Land Registry Services. Ministry of Environment, Lands and Parks, 1997. 4th reprint.

Government of British Columbia

Base Mapping and Geomatic Services Branch. BC Geographical Names Information System

Sessional Papers, Third Session, Twelfth Parliament of the Province of British Columbia. Victoria: King's Printer, 1912.
Dawson, G.H. "Surveyor General's Report", pp. G8-13

Sessional Papers, First Session, Thirteenth Parliament of the Province of British Columbia. Session 1914. Victoria: King's Printer, 1914.
Dawson, G.H. "Report of the Surveyor General", pp. D225-239
Swannell, F.C. "Report on Topographic Surveys in Omineca District", pp. D334-338

Sessional Papers, Second Session, Thirteenth Parliament of the Province of British Columbia. Session 1914. Victoria: King's Printer, 1914.
Dawson, G.H. "Report of the Surveyor General", pp. D297-308
Swannell, F.C. "Exploratory Survey in Omineca District", pp. D351-355

Sessional Papers, Third Session, Thirteenth Parliament of the Province of British Columbia. Session 1915. Victoria: King's Printer, 1915.
Dawson, G.H. "Report of the Surveyor General", pp. D49-61
Swannell, F.C. "Exploration of Finlay and Ingenika Valleys, Cassiar District", pp. D83-90.

Surveyor-General's Department. Field Book Registers, 1908 – 1914

B.C. Archives

B.C. Archives. MS392 – Frank Swannell (journals and correspondence)

B.C. Archives 98002-17 – Frank Swannell (photos and negatives)

Interviews

Sherwood, Jay. Interviews with Gerry Smedley Andrews and Art Swannell – December, 1986 –Audiotape (now on CD)

Websites

"B.C. Archives" – www.bcarchives.gov.bc.ca

"BC Geographical Names" –http://srmwww.gov.bc.ca

83 Mile House 66

100 Mile House 52

105 Mile House 100

Index

141 Mile House 52, 66
153 Mile House 48
Abbott, Edgar 16, 155
Adams, Jack 144
Adams, Mrs. 122
Adolph, Charlie 51-52
Ah Hoo 123
Aitken, Chief Geographer 132
Alaska 38
Alaska Telegraph Line 109
Alberni 43
Albert Lake 156
Alberta 6, 12 (Jasper), 76 (Edmonton), 130, 151-152
Alcohol, see Drinking
Alexander, Jim 90-91, 99-100, 103, 108-110, 112-114, 116-123, 131, 134-138, 141, 144-146, 148-152, 155
Alexander, Mount, see Mount Alexander
Alexis, Tom 111
Allen, Parson 23
Alta Lake 45
American, see United States
Amur tugboat 44
Anderson Lake 27, 46-47, 66
Andrews, Gerry Smedley 26, 87, 90
Anvil Island 45
Archives, BC 1-3
Arctic 54, 96
Ashcroft 14, 17, 23, 63, 92, 100, 105
Astronomer of Canada, Chief 130
Athabasca Landing 151
Automobiles 3, 43, 52, 92, 100-101
Axelgold Peak 119, 156
Axelgold Range 156
Babine 108, 112, 121; see also Fort Babine
Babine, James 97
Babine Lake 32, 37, 110
Babine Trail 107
Bad Luck Mountain 139
Bank, see Northern Crown
Barkerville 123
Bartlett, Mr. 155
Bates Creek 111
Bauer family 45
Bazett, David C. vii
BC Archives 1-3
BC Chief Geographer 132
BC Committee on Geographic Names 156

BC Corporation of Land Surveyors vii
BC Express Co. 52, 65; see also BX
BC Finance Minister 9
BC Forest Service, Branch 26, 95-96, 103-105, 110, 155-156
BC Land Commissioner, Dept., Minister 25, 47, 49, 65, 103, 127
BC Provincial Botanist 131, 152
BC Rail 154
BC Surveyor General 7-9, 16, 23, 25, 31, 33-34, 47, 49-51, 56-57, 65-66, 85-87, 90, 92-93, 97, 100-101, 103-105, 113, 120, 124, 127-132, 134, 147, 152
Bear Lake 39, 70-72, 74-78, 106, 111-112, 116-118, 132, 136, 154
Bear River 78
Bearpaws (snowshoes) 45
Bears 101, 109, 121
Beaton, Mr. 151
Beaver Indians 148, 151
Beaver meadows 100, 114
Beavers 29, 97, 148
Bella Coola 59
Bella Coola Trail 80
Benson, R.M. 19-20, 23, 27, 33, 39, 155
Bentzi Lake 81
Bering Strait 38
Big Bend 138, 140, 145
Big Kettle 113, 118
Big Point 44
Birch, Pat 146
Bishop, R.P. 19-21, 79, 81, 155
Bittancourt's launch 43
Bivouac 67 - 145
Bjornfelt, Mr. 53
Black Canyon 120, 122
Black flies 2, 21, 29
Black, Samuel 131, 141, 146
Blackwater River 23
Blair, William 49
Blanchet, see Mount Blanchet
Blench, Jenny 17
Blench, Tom 17
Boats, see Amur; Steamships
Bobtail Lake 16, 27, 79
Bodeker, Ben and family 144
Bonser, Captain 52-56
Bonser, Mrs. 52
Bower Creek 144, 150

Bower Mountain 144
Bower's Cache 144
Boyd, Mr. 101
Brackendale 45
Braithwaite, John 155
Brandon, Mr. 150
Bridge River 46
Britannia Mine 45
Britannia steamship 45
British, see England
British Columbia, see BC
Bulkley House 38, 72, 108, 111, 117
Bulkley River 12, 132-133
Bulkley Valley 26, 39
Bunting, William 34, 58, 60, 67, 83, 134
Bunting, William (Mrs.) 58, 60, 134
Burden, F.P. 107
Burnet, see Gore, Burnet and Co.
Burns Lake 38, 123-124, 132-133
Butler, William Francis 99, 122-123, 140
Butterfield, Fred 92-93, 107, 123
BX stage line 66, 92
BX sternwheeler 52, 66, 85, 92-93, 107
Cache des Beaux Jours 75
Cache des Bonjours 72
Calgary 152
Cameron, A.W. 52, 66
Canadian Geological Survey 131, 145
Canadian National Railroad 9
Canadian Pacific Railroad (CPR) 6, 14, 23, 63, 91, 105, 130
Canfor 154
Cariboo 7, 14, 27, 48-49, 52, 99-100, 132
Cariboo gold rush 12
Cariboo Observer newspaper 15, 17, 28, 31, 83, 92, 101, 130
Cariboo Road 23, 47-48, 63
Caribou 101, 146, 152
Carrier people 21, 33, 37, 39, 59, 77, 88, 90, 109
Carroll, Dick 133
Cars, see Automobiles
Cartwright, Mr. 151
Cascade Canyon 137, 145, 149
Cascade Range 129
Cassiar 86, 129
Castle Mountain 152
Cataline 20, 38, 41, 123
Catholic church, fathers 17, 34, 36, 39, 58, 72, 74, 76-77, 86, 88-91, 106, 112, 123, 136

Charleson, Jack 18, 21, 32
Charlie, William 81
Charlotte sternwheeler 14-15, 25, 29, 31, 52
Charmer steamship 23
Cheslatta 20, 81, 83
Cheslatta Lake 8, 55, 58, 60, 80-81, 83, 87
Cheslatta Reserves 153
Cheslatta River 83
Cheslatta Trail 80
Chestnut canoe 57
Chicken 124
Chief Astronomer of Canada 130
Chief Charlie Hunter 72, 78, 132-133, 136
Chief, Fort George band 36
Chief Geographer of BC 132
Chief Isidore 54, 67
Chief Louis 55, 60, 81-82
Chila-Chula Rock 23
Chilco steamship 50-56
Chilcotin 28; natives 33
Child, Mr. 122-123
China Rapids 31
Chinese 91-92, 108, 123, 156; see also Nep Yuen
Chinlac 33
Chinlac Rapids 33, 39
Chowsunket Lake 81
Christmas 63
Chronometer 129-130
Chuchi Lake 97
Civil War 6
Clarke, Mr. 21
Clinton 51-52
Cluculz Creek 107
Cluculz Lake 53
Coal 9, 47, 129-130
Coast Range 45
Coccola, Father 36, 39, 72, 74, 76-77, 112, 123
Coccola, Mount, see Mount Coccola
Colley, E.P. 58-59, 83
Collins House 136
Collins Overland Telegraph 12, 38, 106
Collins, storekeeper 18
Columbia River 32
Confederation 6, 9; see also Dominion Day
Connolly, see Fort Connolly
Constantineau, Charles 18, 61, 63, 67
Copley Lake 61, 156
Copley, George Vancouver 3, 26, 33-34, 36-39, 49-50, 52, 54, 56-57, 60-61, 67, 71-72, 77, 79, 90-92, 94-95, 97-98, 101, 103, 105, 107-111, 114, 116-117, 119-120, 122-124, 128, 131-141, 144-146, 148-149, 151-152, 155-156
Copley, Mount, see Mount Copley

Corporation of Land Surveyors vii
Cotton, doctor 68
Cottonwood Canyon 31
Coyotes 138
CPR, see Canadian Pacific
Crooked River 54
Cruise Creek 110
Cullerton, Mr. 39
Currie, Johnnie 94-95, 98, 155
Cust House 151
Cutoff Creek 145
Dams 3; see also Kenney
Danes 121
Danielson, Arne 155
Davidson, John 132, 152, 155
Dawson City 76
Dawson, G.H. 7-8, 128
Decker Lake 124, 133
Deep Creek cabin 98
Deer 36, 54, 101
Delta Creek 146, 148
Deschamps, Bellamy 155
Deschamps, Henry 17-18, 155
Deserter's Canyon 135, 140-142, 150
Devereux, Frank 33, 39
Dick, Captain 23
Dickinson, H.W. 155
Dickinson, Lawrence 101, 155-156
Distributor steamship 40
Diver Lake 113
Diver Mountain 114
Doctors 21, 68
Dogs 22, 113, 115-118, 120, 138, 142, 145-147, 152
Dominion Day celebrations 68, 88-89, 93
Dominion Observatory 130
Dominion Peace River Block 151
Drewry and Twigg (firm) 2
Drewry, W.S. 7
Driftwood Glaciers 111
Driftwood Rapids 151
Driftwood River 36, 38, 71-72, 74-75, 77, 109, 111, 115-117, 154
Driftwood Trail 112
Driftwood Valley 106
Drinking, liquor 39, 43, 123
Driver, Henry 21, 155
Dunkley 151
Dunvegan 151
Echo Lake 119
Echo Mountain 119
Edmonton 76, 151-152
Election, federal 79

Ellis, Sam 155
Ellison, Price 47, 49
Emerald Lakes 141
Endako 59, 63, 133
Endako River 12, 63, 133
Endako Valley 60, 62-63, 124
England, British 50, 58, 142
Enterprise steamer 37, 101, 103, 109
Espee Mountain 138
Euchiniko 79
Europe 7, 38
Eutsuk Lake 58-59
Express, see BC Express
Eyrie triangulation station 19
Fall River 114, 125
Fall River Trail 113
Ferries 11, 20, 55-56, 62-63, 110
Ferris Creek 119
Ferris Mountain 119
Finance Minister, BC 9
Finlay Basin 152
Finlay Forks 121-122, 144, 151, 153
Finlay Junction 125, 136, 144
Finlay Rapids 121, 145, 151
Finlay River 20, 74, 88-89, 120-122, 125, 127, 129-131, 136-147, 149-154, 156
Finlay Valley 125, 145, 152
First Nations, see Native
Fishing 9, 22, 29-31, 37-38, 60, 74, 77-78, 97, 101, 109, 143-144, 148, 152
Fishing Lakes 139, 146
Flies, see Black flies
Forbes, Gilbert 20, 155
Forest Service, see BC Forest Service
Fort Babine 107, 112, 121
Fort Connolly 70, 72, 75-78
Fort Fraser 16-17, 20, 23, 32, 37, 52, 55-56, 58, 63, 67-68, 80, 83, 99, 107, 123, 133-134, 156
Fort Fraser sternwheeler 52-53, 66
Fort George 12, 23, 36, 39, 47-49, 52, 63, 66, 93, 107, 125
Fort George Canyon 23, 25, 31-32, 52, 85
Fort George Navigation and Lumber Co. 32, 52
Fort Grahame 72, 74, 76-78, 116-120, 122, 125, 131-133, 135-136, 140-141, 143, 150-153
Fort McLeod 120, 125, 127-128, 134-136
Fort Nelson 105, 129, 151
Fort St. James 20, 32-36, 38-39, 58, 65, 67-68, 70-72, 74, 76-77, 79, 86-88, 90-97, 99-101, 107-108, 110, 112, 123-125, 127, 130, 134, 148, 151, 153-154, 156
Fort St. John 76, 105, 118, 147-148, 151
Fountain 51

Fox 150
Fox, Mrs. 122
Fox River 143-145, 150, 153
France 150
Francois Lake 58
Frank, Denis 146
Fraser, John A. 15, 27, 31, 57, 65
Fraser Lake 18, 22-23, 25, 32, 38, 56-58, 60, 62-63, 67, 79, 87, 92, 100, 123, 134, 156
Fraser River 3, 11-12, 14-15, 22-23, 32, 37, 50, 52-53, 62-63, 85, 107, 153-154
Freeport 124, 133
French Canadians 18, 20, 144, 150
Fur trade 131, 136
Gauss, Karl 17
Gebhardt, Max 122
Geological Survey of Canada 131, 145
Germansen Creek 125
Germansen Lake 88
Germany 142, 152
Gillis Grave 123
Gillis, Hugh 123
Giscome Portage 50, 53, 125
Glacier House 41
Glaciers 29, 41, 74, 111-112, 118-119, 143, 154
Goat Mountain 119
Goats, see Mountain goats
Gold, Axel (A.M.O.) 103, 107, 110-112, 114-116, 119-120, 155-156; see also Axelgold Peak; Axelgold Range
Gold mining 12, 15, 20; see also Cariboo; Fall River; Germansen; Klondike; Manson; McConnell Creek; Omineca; Slate; Tom; Vital
Goldie Creek 156
Goldie, Hugh 155-156
Goldie, Jack 155-156
Gordon River 43-44
Gore and McGregor 2, 11
Gore, Burnet and Co. 2
Gore, T.S. 2
Grahame, see Fort Grahame
Grand Canyon 54, 56, 58
Grand Rapids 37, 39
Grand Trunk Pacific Railroad 3, 9, 12-13, 23, 28, 40-41, 52, 107, 124, 132-134
Graves, Fort Graham 119; Gillis 123; Graveyard Lake 12, 16; Old Hogem 114; Stikine 148-149
Graveyard Lake 12, 16
Gray, J.H. 13, 22-23, 57, 63
Grease Trail 20, 59
Great Circle trip 56, 58, 153
Greer Creek 156
Greer, Mount, see Mount Greer

Greer, Tom 45, 56-60, 155-156
Greer Valley 67-68
Grizzly bears 109
Groundhog region 47, 128-130, 147, 152
Hallam, William 155
Hallett Creek 60-61
Hallett Lake 60-61, 80-81, 83, 156
Hamilton, Ontario 2
Harrison, fire warden 121
Hazelton 37, 39-41, 110, 112, 147
Hedges, Mr. 150
Hemeyer, Charlie 21, 152, 155
Hemeyer, Charlie (Mrs.) 152
Hetherington, Mr. 155
Hill, John 155
Hills, Dad 21, 107, 155
Hogem, Elmer, see Old Hogem
Holland, Alwin 155
Homestead Act, U.S. 6
Hoo, Ah 123
Horse Meadow Creek 99
Hoy, Bartlett 28
Hoy, Charles 28
Hoy, David 28, 32-33, 39, 54
Huble, Al 151
Huble homestead 50, 53
Hudderle, Mr. 115
Hudson's Bay Co. 2, 6, 15-18, 23, 31-35, 52, 55, 58, 60-61, 65, 67-68, 71, 74, 77, 83, 88, 90, 92, 94, 98, 101, 107-108, 111-113, 116, 118, 120, 123, 131, 134-136, 151, 153-154, 156
Hudson's Hope 146, 151
Hunter, Chief Charlie 72, 78, 132-133, 136
Indata Lake 97, 153
Indians, see Native
Ingenika River 88, 134-135, 137-141, 152, 156
Ingenika Trail 110, 117, 154
Ingenika Valley 152
Inzana Lake 95, 97-100
Isle de Pierre Canyon 23, 26, 32, 51, 54
Isidore, Chief 54, 67
James Island 23
Jasper 12
Jean Marie Creek 156
Johnson, Mr. 54
Johnson's cabin 28
Kains, Tom 7, 86-87
Kamloops 17
Kastberg Creek 154
Kastberg Mountain 156
Kastberg, Viktor 103, 107-110, 114-116, 119, 152, 155-156
Kenney Dam 54, 56, 58, 61, 83, 153

Ketlo, Achille (Ashiel) 71, 155
Ketlo, Thos 67
Ketlo, William 55, 86, 155
Kispiox, Louis 111
Kispiox, Margaret 104-105, 111
Kitwanga 40
Klaskus Lake 79
Klondike 2, 15, 28, 76, 111, 131, 140-141, 143, 145, 151
Klootch 136, 140
Kodak Cascade 145, 149
Kootenay region 2, 7, 9
Kwadacha River 143, 150
La Force, Vital 20, 136
Lac La Hache 1
Laketown 68
Lampitte's Cache 66
Land Commissioner, see BC Land Commissioner
Larson, Louis 123
Larson, Olaf 60-62, 99, 108, 123, 155
Lawrence, Guy 66
Lawrence Lake 156
Leduk 49
Leduke incident 65
Leduke, storekeeper 63, 124
Lesser Slave Lake 152
Liard 150
Liard Pack Trail 143
Lillooet 27, 29, 44, 46, 49-52, 65-66, 79, 92-93
Lillooet River 43-45
Liquor, see Drinking
Long Canyon 144-145, 149
Lookout Mountain 99, 123
Louis, Chief 55, 60, 81-82
Louis' Dumping Ground 58
Lovell Cove 153-154
Luitkart, Mr. 67
Lytton 26, 47
MacAllan, William 68, 71, 74, 79, 134
MacDonald, Archibald 50
MacDonald, Peter 70, 74, 79
Macdougall, Archie 155
MacDougall Creek 156
MacDougall, Douglas 92, 94, 98, 155-156
MacDougall Lake 156
Macfie, John 21, 155
Mackenzie, Alexander 59
Mackenzie River 147
Mahood, Mr. 23
Malzer, Joe 139
Manitoba 2, 6
Manson Creek 37, 88, 96, 98-101, 111-112, 114, 119, 121-123, 125, 140

Manson Creek Trail 123
Mapping 9
Marmots 122
McBride, Richard 9
McConnell Creek rush 121
McConnell Pass 145
McConnell, surveyor 145
McDiarmid, Mr. 130
McDonnell, Pat 155
McDougall, Mr. 20
McElhanney, W.G. 123, 127 (T.A.), 129-130, 134, 152
McGregor, see Gore and McGregor
McIntosh, Mr. 47
McKay, E.G. (E.B.) 25, 47, 49, 65
McKay, Mr. 47
McKennes Road House 132
McLeod, see Fort McLeod
McLeod Lake 135
McMeekin, Jim 155
McPherson, Mr. 107, 112
McVittie's camp 71
Measton, Indian Agent 50
Mehan, General Superintendent 132-133
Mesilinka (Stranger) River 116, 120, 122
Mica Mountain 120
Michel, Albert 94, 97
Middle River 34, 36-37, 72-73, 110, 153
Mile 25 - 117
Mile 280 - 119
Mile 308 - 118
Miller, Del 151
Miller, Slim 53, 107
Milligan, G.B. 86, 105, 123, 129, 151
Milligan, J.M. 98, 123
Milligan, Mount, see Mount Milligan
Milne's Landing 23, 31, 54, 66
Minaret 154
Mineral Monuments 7
Mining 2, 9, 125; see also Gold; Quartz
Mississippi River 6
Mitchell, Jack 95, 97-98, 155-156
Mitchell Pass 156
Mitchell Range 156
Moodie Trail 76, 121-122
Moose 94, 121, 133, 138, 140, 144-145, 147, 152
Morice, Father A.G. 34, 58, 86, 88-91
Moricetown 41
Mosquitoes 2, 11, 21, 29, 45, 114, 137, 142
Mount Alexander 156
Mount Blanchet 72, 91, 153, 156
Mount Coccola 76
Mount Copley 156

Mount Greer 156
Mount Milligan 98, 100
Mount Pope 94-95, 99, 108-109, 123, 153-154
Mount Rognaas 156
Mount Rosetti 156
Mount Sidney Williams 94
Mount Swannell 156
Mount Teegee 112
Mount Trygve 156
Mount William 109
Mount Yuen 117, 156
Mountain goats 119, 141, 144, 152
Mt. see Mount
Mud River 53
Murphy's 66
Murray, A.C. 34, 72, 107, 109, 134
Murray, A.C. (Mrs.) 71
Murray, Annie 71
Murray Lake 83
Murray Mountain 99
Muscovite Lake 122
Muskeg 11, 20, 117, 119, 122
Muskrat 97
Nanaimo 43
Nation Lakes 85, 92-101, 105, 111, 121, 123, 153, 156
Nation River 97-98, 123
Natives 2-3, 5, 12, 14, 16-18, 20-22, 29-31, 33, 36-39, 41, 45-47, 49-52, 55, 57-60, 62-63, 67-74, 76-83, 86-91, 93, 96, 101, 104-107, 111-112, 116, 119, 121-122, 128, 131-136, 140-141, 148, 151-155
Nautley 37-38, 46
Nautley River 23, 38
Nechacco sternwheeler 17, 25-27, 31-33, 53
Nechako Basin 88-89
Nechako Reservoir 58, 153
Nechako River 12-13, 20-23, 26-28, 32-33, 37, 39, 60-62, 65-68, 81, 83, 87, 134, 153, 156
Nechako Road 47, 49, 61, 86, 98
Nechako Settlers Association 47
Nechako Valley 1, 6, 11, 13-23, 25-29, 31-33, 39, 47, 49, 58-59, 63, 65-66, 68, 78, 87, 92-93, 107, 152-153, 156
Nelson, see Fort Nelson
Nep Peak, see Nep Yuen Mountain
Nep Yuen (Jim Young) 5, 91-93, 97, 103, 108, 110-111, 120-124, 131-138, 140-145, 148-149, 150-152, 155-156; see also Yuen
Nep Yuen Mountain (Nep Peak) 117, 156
New Denver 2
New Westminster 132
New Zealand 7

Newport 45
Noakes, A.O. 43, 52, 57
Noonla 54, 67
Norman Lake 156
North Pole 5
North Star, see Polaris
Northern Crown Bank 31, 50, 52, 56, 63, 66, 92
Northland Sun (ship) 152
Northwest Mounted Police 76, 118-120
Northwest Territories 6
Norway, Norse 57
Notch Peak 119
Notch Peak Station 114
Nulki Lake 16, 21
O'Meara, Alf 25-26, 29, 155
Oblates 39
Observatory, Dominion 130
Occidental Hotel 99, 101
Ogston, George 134
Okanagan 9
Old Hogem 113-114
Old Hogem Trail 114
Olson, Anton 28, 57-60, 119, 155
Omineca 20, 87, 94, 105, 111, 115, 122
Omineca gold rush, mining 20, 37, 106, 110-111, 114, 122-123, 125
Omineca Mountains 109-110, 112, 114, 117, 119, 121, 153, 156
Omineca Provincial Park 153
Omineca River 88, 113-114, 117-118, 120, 122
Omineca Trail 112, 117
Ontario 2
Ootsa Lake 55, 58-59
Otterson, Mr. 115, 122-123
Overland Telegraph, see Collins
Owl Creek 29
Pacific Ocean 6, 12, 41, 53, 59
Paddlewheelers, see Steamships, sternwheelers
Parle Pas Rapids 151
Parsnip River 97, 121, 125, 127, 136
Parsnip Valley 125
Patterson, R.M. 120, 131, 137-138, 142, 146, 148-149
Paul Mountain 136
Paul's Branch 143
Peace River 97, 121-122, 125, 136, 145-147, 151
Peace River Block 6, 151; district 86, 105
Peggy Creek 117, 119
Pemberton 43, 45-46
Pemberton Meadows 27, 29, 31, 66
Pemberton Portage 47
Pemberton River 43
Pemberton Valley 29, 45, 47

Petersen, Louis 121-122
Pickard, George 155
Pierre, Aleck 136
Pike, Francis 155
Pinchi 77
Pinchi Point 95
Polaris, North Star 19, 23, 58, 63, 77, 81, 97, 151
Police, see Northwest Mounted Police
Pope, Franklin 106
Pope, Major 109
Pope, Mount, see Mount Pope
Port Renfrew 43-44
Portage 32
Portage Trail 141, 151
Prairie Mountain 144, 150
Prairie Provinces 6, 8, 13, 148
Premier McBride 9
Priests, see Catholic
Priestly 132-133
Prince, Albert 71-72, 79, 155-156
Prince, Benoit 100
Prince, Betsy 110
Prince Edward Island 123
Prince George 7, 48, 52
Prince, Jean Marie 71-73, 79, 155-156
Prince, Maryaz 71-73, 155
Prince Rupert 12, 40-41, 124, 132
Prince Rupert steamship 124, 132
Princess May steamship 105
Provincial Archives 1-3
Provincial Botanist 131, 152
Pulpit Glacier 112
Purvis Lake 96
Quartz Creek 114
Quartz mining 125
Quesnel 11, 14-18, 21-23, 27, 29, 31-32, 34, 52, 56-58, 62-63, 65-66, 79, 92-93, 98-101, 107, 153-154
Quesnel Café 101
Quesnel sternwheeler 52
Railway Belt 6, 128
Railways, trains 3, 6, 9, 12-14, 23, 28, 40-41, 47, 52, 63, 91, 100, 105, 107, 124, 128, 130, 132-134, 152, 154
Rasmussen, P. 122
Red Bluff 112
Reef Canyon 145-146, 149
Renwick, Deputy Minister 49, 65
Reservations 5
Residential school 90
Reveillon 151
Robertson, A.I. 11, 13-14, 16, 19-20, 22, 25, 43, 155

Robinson, Isaac 150
Rocky Mountain Portage 151
Rocky Mountain Trench 144
Rocky Mountains 6, 12, 129
Rognaas Creek 156
Rognaas Lake 61, 156
Rognaas, Mount, see Mount Rognaas
Rognaas, Trygve 47, 49-50, 52-53, 57, 83, 147, 155-156
Rollandot, Mr. 101
Ronayne, John 45
Ronayne, Mrs. 45
Rorison, Mr. 39
Rose Lake 124, 132
Rosette, Mount, see Mount Rosette
Rosette, Sam 103, 108, 110, 112-115, 118, 120, 122-123, 155-156
Ross, factor/postmaster 120, 131, 136, 140
Ross, Lands Minister 65
Rotison Lake 93
Russia 38
Russian American Telegraph 38
Sackner, Mr. 23
Sackrider, H. 155
Saint, see St.
Salmon 29-31, 38, 74, 77
Saltspring Island 43
Saskatchewan 6
Sawridge 152
Schamerhorn, L.H. 155
Schjelderup Creek 156
Schjelderup, Vilhelm 25-26, 29, 33, 39, 43, 45-46, 49, 51-52, 55, 57-58, 60, 65-68, 133, 155-156
Schools, schoolhouse 17, 34, 63, 90, 108, 127, 134, 153-154
Scotland 58
Sekanni people 72, 74, 78, 88, 117, 131-132, 136, 140, 152
Sestino Rapids 66
Seton Lake 27, 29, 46-47, 49-51, 66
Sharkey, Patrick 22-23, 27-28, 65-67, 79, 98, 155
Sharkey, Tom 92
Shass Mountain 109
Shepherd's Road House 63
Ships, see Steamships, sternwheelers
Short Portage 50-52, 66
Shovel Creek 124
Sidney Island 43
Sifton Pass 144
Sinkut Astro pier 130
Siwash 3, 22, 39, 60, 77, 93, 97, 113, 148
Skeena River 3, 12, 32, 39-40, 70
Skins Spillway 54

Slate Creek Placer Mine 114-115, 122
Slave Lake, Lesser 152
Smallpox 12
Smith, Mr. 23, 152
Smithers 124, 132
Snow 23
Soda Creek 14-15, 22-23, 29, 52, 66, 92-93, 107, 154
Soda Creek Hotel 29
Souris, PEI 123
Souses, F. 51
Spetch, Sid 155
St. Joseph's Residential School 90
St. Thomas Bay 58
Stagecoaches 2
Standing Rock Canyon 67
Stanwell-Fletcher, Theodora 106, 112
Steamships, sternwheelers 2, 14-15, 17, 23, 25-26, 29, 31-32, 37, 40-41, 45-46, 50-56, 58, 92, 100-101, 103, 105, 107, 124-125, 132, 151-154
Steele, Billy 122-123
Stella 55, 67, 71, 133
Stellako Reserve 22
Stephens, Frank 93, 96, 110, 112
Stephens, Frank (Mrs.) 112
Stephens River 101
Sternwheelers, see Steamships
Stewart Lake 92
Stikine people 148-149
Stikine River 146
Stila-chula 81
Stoney Creek 16 (telegraph cabin), 17-18, 21-22, 39, 60, 66-67, 107
Stranger River, see Mesilinka
Stuart Lake 29-33, 36-37, 39, 49, 71, 92 (Stewart), 94-95, 100-101, 104, 109, 144, 153-154
Stuart River 21, 23, 33, 36-39, 71, 153
Studebaker 43, 100
Sturgess, A. 155
Sullivan, Ed 122
Summit Lake 54
Summit Lake (Whistler) 45
Summit Moody Trail, see Moody Trail
Surveyor-General, see BC Surveyor General
Sustut Provincial Park 153
Sustut River 74
Sutton, Father 123
Sutton, Mr. 107
Swannell, Ada 45-47, 51, 105, 107, 132, 150
Swannell, Frank, see references throughout
Swannell, Lorne 132
Swannell, Mount, see Mount Swannell

Swannell Ranges 139, 156
Swannell River 137, 156
Swedes 22, 43, 53
Swinerton, Mr. 23, 155
Tache Point 95, 109
Tache River 34, 36-37, 39, 71, 94, 109
Takla Lake 5, 34, 36-38, 72, 77, 79, 87, 90-96, 100-101, 106-107, 110-115, 117, 153-154, 156
Takla Landing 37, 96
Takla Narrows 111
Tatchi River 101
Tatla Lake 36, 72, 95-96, 101, 110, 112, 117
Tatlow, R.G. 9
Ta-wat-enslinstay Rapids 109
Taylor, T.H. 47, 129, 131, 147, 152
Taylor, W.A. 6-7
Tchentlo Lake 96-97
Teegee Creek 112
Teegee, Daniel 77, 87, 107, 112
Teegee, Mount, see Mount Teegee
Teepees 148, 151
Telegraph 2, 11-12, 15-16, 20, 22, 27, 31, 38-39, 41, 63, 66, 76, 79, 106, 109, 130-131, 147
Telegraph Creek 22
Terminal Steamship line 45
Tetachuk Lake 59
Tetachuk River 59
Theodolite 5
Thomas, Seymour 81
Thompson, Don 1
Thompson, Norman 155-156
Thompson, Roy 155
Thorpe, Joe 68, 155
Thucatade River 146
Thutade Lake 131, 139, 147-148
Titanic 58
Tod, Donald 67, 123
Tod, George 35
Tom Creek 111, 125
Too-Tizzi Lake 115, 118-120
Too-Tizzi Mountain 120
Toronto 2
Transit 5
Traveller's Hotel 124
Tremblai, see Trembleur
Tremblay, see Trembleur
Trembleur Lake 34, 36-37, 39, 71, 73, 77, 79, 94-95, 100-101, 103, 109-110, 153
Triangle Lake 81
Triangulation 7
Trout Creek 123
Trygve Creek 156
Trygve Lake 156

Trygve, Mount, see Mount Trygve
Tsayta Lake 96-97, 153
Tsinkut 21
Tsinkut Crossing 108
Tsinkut Falls 13
Tutachi Lake 138
Tu-wat-en-indlay Rapids 116-117
Tweedsmuir Park 156
Twigg, see Drewry and Twigg
Tyee, Skin 81
United States, American 6, 23, 32, 38, 45
University of Alberta, Edmonton 152
University of Toronto 2
Van Arsdol, Chief Engineer 133
Vancouver 14, 23, 26, 45, 47, 60, 63, 65, 92, 105, 124, 132, 146, 152
Vancouver Island 8
Vanderhoof 21, 23, 130, 153, 156
Victoria 2, 14, 23, 25-26, 39, 41, 43, 47, 50-52, 56, 63, 65, 79, 83, 92-93, 100-101, 105, 107, 110, 124, 134, 136, 150-152
Vital Creek 20, 125
Vital's Bar 20, 136
Vital's Ferry 20, 55
War 6 (Civil War), 142, 150, 152, 156
Warner, Mr. 150
Waters Camp 50
Waters, W.E. 50, 60, 155
Welch, Percy 67, 155
West Landing 37, 104, 107, 110-112
Whistler area 45
White Bluff 112
White, George 67, 155
Whitesail Lake 55-56, 58
Wilkie, W. 31, 155
Wilkins, G. 20, 23
William, Mount, see Mount William
Williams Lake 90-91
Williams, Mount, see Mount Sidney Williams
Williams, Sidney 77
Williston Lake 151
Williston Reservoir 153
Wilton, fire warden 121
Windlass Portage 135
Wolverine Range 121
Wolverine Summit 122
World War 142, 150, 152, 156
Wrede Creek 138
Wright, Mr. 47
Wyllie, Mr. 155
Yellowhead Pass 12
Young, Jack 44
Young, Jim, see Nep Yuen

Yuen Creek 156
Yuen Lake 156
Yuen, Nep, see Nep Yuen; Mount Yuen; Yuen Creek; Yuen Lake
Yukon 2
Yukon River 32
Yukon Telegraph Trail 2, 11, 15-16, 20, 27, 31, 39, 41, 63, 66, 76, 79, 130-131, 147